SO-AFU-682

"A great introduction to machine learning from a world-class practitioner."
— **Karolis Urbonas**, Head of Data Science at Amazon

"I wish such a book existed when I was a statistics graduate student trying to learn about machine learning."
— **Chao Han**, VP, Head of R&D at Lucidworks

"Andriy's book does a fantastic job of cutting the noise and hitting the tracks and full speed from the first page."
— **Sujeet Varakhedi**, Head of Engineering at eBay

"A wonderful book for engineers who want to incorporate ML in their day-to-day work without necessarily spending an enormous amount of time."
— **Deepak Agarwal**, VP of Artificial Intelligence at LinkedIn

"Excellent read to get started with Machine Learning."
— **Vincent Pollet**, Head of Research at Nuance

with back cover text from **Peter Norvig** and **Aurélien Geron**

The Hundred-Page Machine Learning Book

Andriy Burkov

Copyright ©2019 Andriy Burkov

All rights reserved. This book is distributed on the "read first, buy later" principle. The latter implies that anyone can obtain a copy of the book by any means available, read it and share it with anyone else. However, if you read the book, liked it or found it helpful or useful in any way, you have to buy it. For further information, please email author@themlbook.com.

ISBN 978-1-9995795-0-0

Publisher: Andriy Burkov

To my parents:
Tatiana and Valeriy

and to my family:
daughters Catherine and Eva,
and brother Dmitriy

"All models are wrong, but some are useful."
— *George Box*

"If I had more time, I would have written a shorter letter."
— *Blaise Pascal*

The book is distributed on the "read first, buy later" principle.

Contents

Foreword

The last twenty years have witnessed an explosion in the availability of enormous quantities of data and, correspondingly, of interest in statistical and machine learning applications. The impact has been profound. Ten years ago, when I was able to attract a full class of MBA students to my new statistical learning elective, my colleagues were astonished because our department struggled to fill most electives. Today we offer a Master's in Business Analytics, which is the largest specialized master's program in the school and has application volume rivaling those of our MBA programs. Our course offerings have increased dramatically, yet our students still complain that the classes are all full. Our experience is not unique, with data science and machine learning programs springing up at an extraordinary rate as the demand for individuals trained in this area has blossomed.

This demand is driven by a simple, but undeniable, fact. Machine learning approaches have produced significant new insights in numerous settings such as the social sciences, business, biology and medicine, to name just a few. As a result, there is a tremendous demand for individuals with the requisite skill set. However, training students in these skills has been challenging because most of the early literature on these methods was aimed at academics and concentrated on statistical and theoretical properties of the fitting algorithms or resulting estimators. There was little support for researchers and practitioners who needed help in implementing a given method on real-world problems. These individuals needed to understand the range of methods that can be applied to each problem, along with their assumptions, strengths and weaknesses. But theoretical properties or detailed information on the fitting algorithms were far less important. Our goal when we wrote "An Introduction to Statistical Learning with R" (ISLR) was to provide a resource for this group. The enthusiasm with which it was received demonstrates the demand that exists within the community.

"The Hundred-Page Machine Learning Book" follows a similar paradigm. As with ISLR, it skips involved theoretical derivations in favor of providing the reader with key details on how to implement the various approaches. This is a compact "how to do data science" manual and I predict it will become a go to resource for academics and practitioners alike. At 100 pages (or a little more), the book is short enough to read in a single sitting. Yet, despite its length, it covers all the major machine learning approaches, ranging from classical linear and logistic regression, through to modern support vector machines, deep learning, boosting, and random forests. There is also no shortage of details on the various approaches and the interested reader can gain further information on any particular method via the innovative companion book wiki. The book does not assume any high level mathematical or statistical training, or even programming experience, so should be accessible to almost anyone willing to invest the time to learn about these methods. It should certainly be required reading for anyone starting a PhD program in this area and will serve as a useful reference as they progress further. Finally, the book illustrates some of the algorithms using Python code, one of the most popular coding languages for machine learning. I would highly recommend "The Hundred-Page Machine Learning Book" for both the beginner looking to learn more about machine learning, and the experienced practitioner seeking to extend their knowledge base.

Gareth James, Professor of Data Sciences and Operations at University of Southern California, co-author (with Witten, Hastie and Tibshirani), of the best-selling book ***An Introduction to Statistical Learning, with Applications in R***

Preface

Let's start by telling the truth: machines don't learn. What a typical "learning machine" does, is finding a mathematical formula, which, when applied to a collection of inputs (called "training data"), produces the desired outputs. This mathematical formula also generates the correct outputs for most other inputs (distinct from the training data) on the condition that those inputs come from the same or a similar statistical distribution as the one the training data was drawn from.

Why isn't that learning? Because if you slightly distort the inputs, the output is very likely to become completely wrong. It's not how learning in animals works. If you learned to play a video game by looking straight at the screen, you would still be a good player if someone rotates the screen slightly. A machine learning algorithm, if it was trained by "looking" straight at the screen, unless it was also trained to recognize rotation, will fail to play the game on a rotated screen.

So why the name "machine learning" then? The reason, as is often the case, is marketing: Arthur Samuel, an American pioneer in the field of computer gaming and artificial intelligence, coined the term in 1959 while at IBM. Similarly to how in the 2010s IBM tried to market the term "cognitive computing" to stand out from competition, in the 1960s, IBM used the new cool term "machine learning" to attract both clients and talented employees.

As you can see, just like artificial intelligence is not intelligence, machine learning is not learning. However, machine learning is a universally recognized term that usually refers to the science and engineering of building machines capable of doing various useful things without being explicitly programmed to do so. So, the word "learning" in the term is used by analogy with the learning in animals rather than literally.

Who This Book is For

This book contains only those parts of the vast body of material on machine learning developed since the 1960s that have proven to have a significant practical value. A beginner in machine learning will find in this book just enough details to get a comfortable level of understanding of the field and start asking the right questions.

Practitioners with experience can use this book as a collection of directions for further self-improvement. The book also comes in handy when brainstorming at the beginning of a project, when you try to answer the question whether a given technical or business problem is "machine-learnable" and, if yes, which techniques you should try to solve it.

How to Use This Book

If you are about to start learning machine learning, you should read this book from the beginning to the end. (It's just a hundred pages, not a big deal.) If you are interested in a specific topic covered in the book and want to know more, most sections have a QR code.

QR Code

By scanning one of those QR codes with your phone, you will get a link to a page on the book's companion wiki theMLbook.com with additional materials: recommended reads, videos, Q&As, code snippets, tutorials, and other bonuses. The book's wiki is continuously updated with contributions from the book's author himself as well as volunteers from all over the world. So this book, like a good wine, keeps getting better after you buy it.

Scan the QR code on the left to get to the book's wiki.

Some sections don't have a QR code, but they still most likely have a wiki page. You can find it by submitting the section's title to the wiki's search engine.

Should You Buy This Book?

This book is distributed on the "read first, buy later" principle. I firmly believe that paying for the content before consuming it is buying a pig in a poke. You can see and try a car in a dealership before you buy it. You can try on a shirt or a dress in a department store. You have to be able to read a book before paying for it.

The *read first, buy later* principle implies that you can freely download the book, read it and share it with your friends and colleagues. If you read and liked the book, or found it helpful or useful in your work, business or studies, then buy it.

Now you are all set. Enjoy your reading!

Andriy Burkov

Chapter 1

Introduction

1.1 What is Machine Learning

Machine learning is a subfield of computer science that is concerned with building algorithms which, to be useful, rely on a collection of examples of some phenomenon. These examples can come from nature, be handcrafted by humans or generated by another algorithm.

Machine learning can also be defined as the process of solving a practical problem by 1) gathering a dataset, and 2) algorithmically building a statistical model based on that dataset. That statistical model is assumed to be used somehow to solve the practical problem.

To save keystrokes, I use the terms "learning" and "machine learning" interchangeably.

1.2 Types of Learning

Learning can be supervised, semi-supervised, unsupervised and reinforcement.

1.2.1 Supervised Learning

In **supervised learning**[1], the **dataset** is the collection of **labeled examples** $\{(\mathbf{x}_i, y_i)\}_{i=1}^{N}$. Each element $\mathbf{x}_i$ among N is called a **feature vector**. A feature vector is a vector in which each dimension $j = 1, \ldots, D$ contains a value that describes the example somehow. That value is called a **feature** and is denoted as $x^{(j)}$. For instance, if each example $\mathbf{x}$ in our collection represents a person, then the first feature, $x^{(1)}$, could contain height in cm, the

---

[1]If a term is **in bold**, that means that the term can be found in the index at the end of the book.

second feature, $x^{(2)}$, could contain weight in kg, $x^{(3)}$ could contain gender, and so on. For all examples in the dataset, the feature at position j in the feature vector always contains the same kind of information. It means that if $x_i^{(2)}$ contains weight in kg in some example $\mathbf{x}_i$, then $x_k^{(2)}$ will also contain weight in kg in every example $\mathbf{x}_k$, $k = 1, \ldots, N$. The **label** y_i can be either an element belonging to a finite set of **classes** $\{1, 2, \ldots, C\}$, or a real number, or a more complex structure, like a vector, a matrix, a tree, or a graph. Unless otherwise stated, in this book y_i is either one of a finite set of classes or a real number[2]. You can see a class as a category to which an example belongs. For instance, if your examples are email messages and your problem is spam detection, then you have two classes $\{spam, not\_spam\}$.

The goal of a **supervised learning algorithm** is to use the dataset to produce a **model** that takes a feature vector $\mathbf{x}$ as input and outputs information that allows deducing the label for this feature vector. For instance, the model created using the dataset of people could take as input a feature vector describing a person and output a probability that the person has cancer.

1.2.2 Unsupervised Learning

In **unsupervised learning**, the dataset is a collection of **unlabeled examples** $\{\mathbf{x}_i\}_{i=1}^N$. Again, $\mathbf{x}$ is a feature vector, and the goal of an **unsupervised learning algorithm** is to create a **model** that takes a feature vector $\mathbf{x}$ as input and either transforms it into another vector or into a value that can be used to solve a practical problem. For example, in **clustering**, the model returns the id of the cluster for each feature vector in the dataset. In **dimensionality reduction**, the output of the model is a feature vector that has fewer features than the input $\mathbf{x}$; in **outlier detection**, the output is a real number that indicates how $\mathbf{x}$ is different from a "typical" example in the dataset.

1.2.3 Semi-Supervised Learning

In **semi-supervised learning**, the dataset contains both labeled and unlabeled examples. Usually, the quantity of unlabeled examples is much higher than the number of labeled examples. The goal of a **semi-supervised learning algorithm** is the same as the goal of the supervised learning algorithm. The hope here is that using many unlabeled examples can help the learning algorithm to find (we might say "produce" or "compute") a better model.

It could look counter-intuitive that learning could benefit from adding more unlabeled examples. It seems like we add more uncertainty to the problem. However, when you add unlabeled examples, you add more information about your problem: a larger sample reflects better the probability distribution the data we labeled came from. Theoretically, a learning algorithm should be able to leverage this additional information.

---

[2]A real number is a quantity that can represent a distance along a line. Examples: $0, -256.34, 1000, 1000.2$.

1.2.4 Reinforcement Learning

Reinforcement learning is a subfield of machine learning where the machine "lives" in an environment and is capable of perceiving the *state* of that environment as a vector of features. The machine can execute *actions* in every state. Different actions bring different *rewards* and could also move the machine to another state of the environment. The goal of a reinforcement learning algorithm is to learn a *policy*.

 A policy is a function (similar to the model in supervised learning) that takes the feature vector of a state as input and outputs an optimal action to execute in that state. The action is optimal if it maximizes the *expected average reward*.

Reinforcement learning solves a particular kind of problem where decision making is sequential, and the goal is long-term, such as game playing, robotics, resource management, or logistics. In this book, I put emphasis on one-shot decision making where input examples are independent of one another and the predictions made in the past. I leave reinforcement learning out of the scope of this book.

1.3 How Supervised Learning Works

In this section, I briefly explain how supervised learning works so that you have the picture of the whole process before we go into detail. I decided to use supervised learning as an example because it's the type of machine learning most frequently used in practice.

The supervised learning process starts with gathering the data. The data for supervised learning is a collection of pairs (input, output). Input could be anything, for example, email messages, pictures, or sensor measurements. Outputs are usually real numbers, or labels (e.g. "spam", "not_spam", "cat", "dog", "mouse", etc). In some cases, outputs are vectors (e.g., four coordinates of the rectangle around a person on the picture), sequences (e.g. ["adjective", "adjective", "noun"] for the input "big beautiful car"), or have some other structure.

Let's say the problem that you want to solve using supervised learning is spam detection. You gather the data, for example, 10,000 email messages, each with a label either "spam" or "not_spam" (you could add those labels manually or pay someone to do that for you). Now, you have to convert each email message into a feature vector.

The data analyst decides, based on their experience, how to convert a real-world entity, such as an email message, into a feature vector. One common way to convert a text into a feature vector, called **bag of words**, is to take a dictionary of English words (let's say it contains 20,000 alphabetically sorted words) and stipulate that in our feature vector:

- the first feature is equal to 1 if the email message contains the word "a"; otherwise, this feature is 0;

- the second feature is equal to 1 if the email message contains the word "aaron"; otherwise, this feature equals 0;
- ...
- the feature at position 20,000 is equal to 1 if the email message contains the word "zulu"; otherwise, this feature is equal to 0.

You repeat the above procedure for every email message in our collection, which gives us 10,000 feature vectors (each vector having the dimensionality of 20,000) and a label ("spam"/"not_spam").

Now you have machine-readable input data, but the output labels are still in the form of human-readable text. Some learning algorithms require transforming labels into numbers. For example, some algorithms require numbers like 0 (to represent the label "not_spam") and 1 (to represent the label "spam"). The algorithm I use to illustrate supervised learning is called **Support Vector Machine** (SVM). This algorithm requires that the positive label (in our case it's "spam") has the numeric value of $+1$ (one), and the negative label ("not_spam") has the value of -1 (minus one).

At this point, you have a **dataset** and a **learning algorithm**, so you are ready to apply the learning algorithm to the dataset to get the **model**.

SVM sees every feature vector as a point in a high-dimensional space (in our case, space is 20,000-dimensional). The algorithm puts all feature vectors on an imaginary 20,000-dimensional plot and draws an imaginary 19,999-dimensional line (a *hyperplane*) that separates examples with positive labels from examples with negative labels. In machine learning, the boundary separating the examples of different classes is called the **decision boundary**.

The equation of the hyperplane is given by two **parameters**, a real-valued vector $\mathbf{w}$ of the same dimensionality as our input feature vector $\mathbf{x}$, and a real number b like this:

$$\mathbf{w}\mathbf{x} - b = 0,$$

where the expression $\mathbf{w}\mathbf{x}$ means $w^{(1)}x^{(1)} + w^{(2)}x^{(2)} + \ldots + w^{(D)}x^{(D)}$, and D is the number of dimensions of the feature vector $\mathbf{x}$.

(If some equations aren't clear to you right now, in Chapter 2 we revisit the math and statistical concepts necessary to understand them. For the moment, try to get an intuition of what's happening here. It all becomes more clear after you read the next chapter.)

Now, the predicted label for some input feature vector $\mathbf{x}$ is given like this:

$$y = \text{sign}(\mathbf{w}\mathbf{x} - b),$$

where sign is a mathematical operator that takes any value as input and returns $+1$ if the input is a positive number or -1 if the input is a negative number.

The goal of the learning algorithm — SVM in this case — is to leverage the dataset and find the optimal values $\mathbf{w}^*$ and b^* for parameters $\mathbf{w}$ and b. Once the learning algorithm identifies these optimal values, the **model** $f(\mathbf{x})$ is then defined as:

$$f(\mathbf{x}) = \text{sign}(\mathbf{w}^*\mathbf{x} - b^*)$$

Therefore, to predict whether an email message is spam or not spam using an SVM model, you have to take the text of the message, convert it into a feature vector, then multiply this vector by $\mathbf{w}^*$, subtract b^* and take the sign of the result. This will give us the prediction (+1 means "spam", −1 means "not_spam").

Now, how does the machine find $\mathbf{w}^*$ and b^*? It solves an optimization problem. Machines are good at optimizing functions under constraints.

So what are the constraints we want to satisfy here? First of all, we want the model to predict the labels of our 10,000 examples correctly. Remember that each example $i = 1, \ldots, 10000$ is given by a pair $(\mathbf{x}_i, y_i)$, where $\mathbf{x}_i$ is the feature vector of example i and y_i is its label that takes values either −1 or +1. So the constraints are naturally:

$$\mathbf{w}\mathbf{x}_i - b \geq +1 \quad \text{if } y_i = +1,$$
$$\mathbf{w}\mathbf{x}_i - b \leq -1 \quad \text{if } y_i = -1.$$

We would also prefer that the hyperplane separates positive examples from negative ones with the largest **margin**. The margin is the distance between the closest examples of two classes, as defined by the decision boundary. A large margin contributes to a better **generalization**, that is how well the model will classify new examples in the future. To achieve that, we need to minimize the Euclidean norm of $\mathbf{w}$ denoted by $\|\mathbf{w}\|$ and given by $\sqrt{\sum_{j=1}^{D}(w^{(j)})^2}$.

So, the optimization problem that we want the machine to solve looks like this:

Minimize $\|\mathbf{w}\|$ *subject to* $y_i(\mathbf{w}\mathbf{x}_i - b) \geq 1$ *for* $i = 1, \ldots, N$. The expression $y_i(\mathbf{w}\mathbf{x}_i - b) \geq 1$ is just a compact way to write the above two constraints.

The solution of this optimization problem, given by $\mathbf{w}^*$ and b^*, is called the **statistical model**, or, simply, the **model**. The process of building the model is called **training**.

For two-dimensional feature vectors, the problem and the solution can be visualized as shown in Figure 1.1. The blue and orange circles represent, respectively, positive and negative examples, and the line given by $\mathbf{w}\mathbf{x} - b = 0$ is the decision boundary.

Why, by minimizing the norm of $\mathbf{w}$, do we find the highest margin between the two classes? Geometrically, the equations $\mathbf{w}\mathbf{x} - b = 1$ and $\mathbf{w}\mathbf{x} - b = -1$ define two parallel hyperplanes, as you see in Figure 1.1. The distance between these hyperplanes is given by $\frac{2}{\|\mathbf{w}\|}$, so the smaller the norm $\|\mathbf{w}\|$, the larger the distance between these two hyperplanes.

That's how Support Vector Machines work. This particular version of the algorithm builds the so-called *linear model*. It's called linear because the decision boundary is a straight line

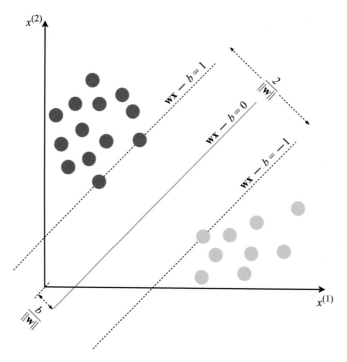

Figure 1.1: An example of an SVM model for two-dimensional feature vectors.

(or a plane, or a hyperplane). SVM can also incorporate **kernels** that can make the decision boundary arbitrarily non-linear. In some cases, it could be impossible to perfectly separate the two groups of points because of noise in the data, errors of labeling, or **outliers** (examples very different from a "typical" example in the dataset). Another version of SVM can also incorporate a penalty hyperparameter[3] for misclassification of training examples of specific classes. We study the SVM algorithm in more detail in Chapter 3.

At this point, you should retain the following: any classification learning algorithm that builds a model implicitly or explicitly creates a decision boundary. The decision boundary can be straight, or curved, or it can have a complex form, or it can be a superposition of some geometrical figures. The form of the decision boundary determines the **accuracy** of the model (that is the ratio of examples whose labels are predicted correctly). The form of the decision boundary, the way it is algorithmically or mathematically computed based on the training data, differentiates one learning algorithm from another.

In practice, there are two other essential differentiators of learning algorithms to consider: speed of model building and prediction processing time. In many practical cases, you would

---

[3]A hyperparameter is a property of a learning algorithm, usually (but not always) having a numerical value. That value influences the way the algorithm works. Those values aren't learned by the algorithm itself from data. They have to be set by the data analyst before running the algorithm.

prefer a learning algorithm that builds a less accurate model quickly. Additionally, you might prefer a less accurate model that is much quicker at making predictions.

1.4 Why the Model Works on New Data

Why is a machine-learned model capable of predicting correctly the labels of new, previously unseen examples? To understand that, look at the plot in Figure 1.1. If two classes are separable from one another by a decision boundary, then, obviously, examples that belong to each class are located in two different subspaces which the decision boundary creates.

If the examples used for training were selected randomly, independently of one another, and following the same procedure, then, statistically, it is *more likely* that the new negative example will be located on the plot somewhere not too far from other negative examples. The same concerns the new positive example: it will *likely* come from the surroundings of other positive examples. In such a case, our decision boundary will still, *with high probability*, separate well new positive and negative examples from one another. For other, *less likely situations*, our model will make errors, but because such situations are less likely, the number of errors will likely be smaller than the number of correct predictions.

Intuitively, the larger is the set of training examples, the more unlikely that the new examples will be dissimilar to (and lie on the plot far from) the examples used for training.

To minimize the probability of making errors on new examples, the SVM algorithm, by looking for the largest margin, explicitly tries to draw the decision boundary in such a way that it lies as far as possible from examples of both classes.

The reader interested in knowing more about the *learnability* and understanding the close relationship between the model error, the size of the training set, the form of the mathematical equation that defines the model, and the time it takes to build the model is encouraged to read about the *PAC learning*. The PAC (for "probably approximately correct") learning theory helps to analyze whether and under what conditions a learning algorithm will probably output an approximately correct classifier.

Chapter 2

Notation and Definitions

2.1 Notation

Let's start by revisiting the mathematical notation we all learned at school, but some likely forgot right after the prom.

2.1.1 Data Structures

A **scalar** is a simple numerical value, like 15 or -3.25. Variables or constants that take scalar values are denoted by an italic letter, like x or a.

A **vector** is an ordered list of scalar values, called attributes. We denote a vector as a bold character, for example, $\mathbf{x}$ or $\mathbf{w}$. Vectors can be visualized as arrows that point to some directions as well as points in a multi-dimensional space. Illustrations of three two-dimensional vectors, $\mathbf{a} = [2, 3]$, $\mathbf{b} = [-2, 5]$, and $\mathbf{c} = [1, 0]$ are given in Figure 2.1. We denote an attribute of a vector as an italic value with an index, like this: $w^{(j)}$ or $x^{(j)}$. The index j denotes a specific **dimension** of the vector, the position of an attribute in the list. For instance, in the vector $\mathbf{a}$ shown in red in Figure 2.1, $a^{(1)} = 2$ and $a^{(2)} = 3$.

The notation $x^{(j)}$ should not be confused with the power operator, such as the 2 in x^2 (squared) or 3 in x^3 (cubed). If we want to apply a power operator, say square, to an indexed attribute of a vector, we write like this: $(x^{(j)})^2$.

A variable can have two or more indices, like this: $x_i^{(j)}$ or like this $x_{i,j}^{(k)}$. For example, in neural networks, we denote as $x_{l,u}^{(j)}$ the input feature j of unit u in layer l.

A **matrix** is a rectangular array of numbers arranged in rows and columns. Below is an example of a matrix with two rows and three columns,

$$\begin{bmatrix} 2 & 4 & -3 \\ 21 & -6 & -1 \end{bmatrix}.$$

Matrices are denoted with bold capital letters, such as $\mathbf{A}$ or $\mathbf{W}$.

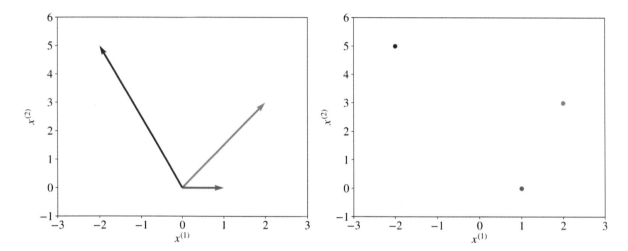

Figure 2.1: Three vectors visualized as directions and as points.

A **set** is an unordered collection of unique elements. We denote a set as a calligraphic capital character, for example, $\mathcal{S}$. A set of numbers can be finite (include a fixed amount of values). In this case, it is denoted using accolades, for example, $\{1, 3, 18, 23, 235\}$ or $\{x_1, x_2, x_3, x_4, \ldots, x_n\}$. A set can be infinite and include all values in some interval. If a set includes all values between a and b, including a and b, it is denoted using brackets as $[a, b]$. If the set doesn't include the values a and b, such a set is denoted using parentheses like this: (a, b). For example, the set $[0, 1]$ includes such values as 0, 0.0001, 0.25, 0.784, 0.9995, and 1.0. A special set denoted $\mathbb{R}$ includes all numbers from minus infinity to plus infinity.

When an element x belongs to a set $\mathcal{S}$, we write $x \in \mathcal{S}$. We can obtain a new set $\mathcal{S}_3$ as an **intersection** of two sets $\mathcal{S}_1$ and $\mathcal{S}_2$. In this case, we write $\mathcal{S}_3 \leftarrow \mathcal{S}_1 \cap \mathcal{S}_2$. For example $\{1, 3, 5, 8\} \cap \{1, 8, 4\}$ gives the new set $\{1, 8\}$.

We can obtain a new set $\mathcal{S}_3$ as a *union* of two sets $\mathcal{S}_1$ and $\mathcal{S}_2$. In this case, we write $\mathcal{S}_3 \leftarrow \mathcal{S}_1 \cup \mathcal{S}_2$. For example $\{1, 3, 5, 8\} \cup \{1, 8, 4\}$ gives the new set $\{1, 3, 4, 5, 8\}$.

2.1.2 Capital Sigma Notation

The summation over a collection $X = \{x_1, x_2, \ldots, x_{n-1}, x_n\}$ or over the attributes of a vector $\mathbf{x} = [x^{(1)}, x^{(2)}, \ldots, x^{(m-1)}, x^{(m)}]$ is denoted like this:

$$\sum_{i=1}^{n} x_i \overset{\text{def}}{=} x_1 + x_2 + \ldots + x_{n-1} + x_n, \text{ or else: } \sum_{j=1}^{m} x^{(j)} \overset{\text{def}}{=} x^{(1)} + x^{(2)} + \ldots + x^{(m-1)} + x^{(m)}.$$

The notation $\overset{\text{def}}{=}$ means "is defined as".

2.1.3 Capital Pi Notation

A notation analogous to capital sigma is the **capital pi notation.** It denotes a product of elements in a collection or attributes of a vector:

$$\prod_{i=1}^{n} x_i \overset{\text{def}}{=} x_1 \cdot x_2 \cdot \ldots \cdot x_{n-1} \cdot x_n,$$

where $a \cdot b$ means a multiplied by b. Where possible, we omit $\cdot$ to simplify the notation, so ab also means a multiplied by b.

2.1.4 Operations on Sets

A derived set creation operator looks like this: $\mathcal{S}' \leftarrow \{x^2 \mid x \in \mathcal{S}, x > 3\}$. This notation means that we create a new set $\mathcal{S}'$ by putting into it x squared such that x is in $\mathcal{S}$, and x is greater than 3.

The cardinality operator $|\mathcal{S}|$ returns the number of elements in set $\mathcal{S}$.

2.1.5 Operations on Vectors

The sum of two vectors $\mathbf{x} + \mathbf{z}$ is defined as the vector $[x^{(1)} + z^{(1)}, x^{(2)} + z^{(2)}, \ldots, x^{(m)} + z^{(m)}]$.

The difference of two vectors $\mathbf{x} - \mathbf{z}$ is defined as $[x^{(1)} - z^{(1)}, x^{(2)} - z^{(2)}, \ldots, x^{(m)} - z^{(m)}]$.

A vector multiplied by a scalar is a vector. For example $\mathbf{x}c \overset{\text{def}}{=} [cx^{(1)}, cx^{(2)}, \ldots, cx^{(m)}]$.

A **dot-product** of two vectors is a scalar. For example, $\mathbf{wx} \overset{\text{def}}{=} \sum_{i=1}^{m} w^{(i)} x^{(i)}$. In some books, the dot-product is denoted as $\mathbf{w} \cdot \mathbf{x}$. The two vectors must be of the same dimensionality. Otherwise, the dot-product is undefined.

The multiplication of a matrix $\mathbf{W}$ by a vector $\mathbf{x}$ results in another vector. Let our matrix be,

$$\mathbf{W} = \begin{bmatrix} w^{(1,1)} & w^{(1,2)} & w^{(1,3)} \\ w^{(2,1)} & w^{(2,2)} & w^{(2,3)} \end{bmatrix}.$$

When vectors participate in operations on matrices, a vector is by default represented as a matrix with one column. When the vector is on the right of the matrix, it remains a column vector. We can only multiply a matrix by vector if the vector has the same number of rows as the number of columns in the matrix. Let our vector be $\mathbf{x} \overset{\text{def}}{=} [x^{(1)}, x^{(2)}, x^{(3)}]$. Then $\mathbf{Wx}$ is a two-dimensional vector defined as,

$$
\begin{aligned}
\mathbf{Wx} &= \begin{bmatrix} w^{(1,1)} & w^{(1,2)} & w^{(1,3)} \\ w^{(2,1)} & w^{(2,2)} & w^{(2,3)} \end{bmatrix} \begin{bmatrix} x^{(1)} \\ x^{(2)} \\ x^{(3)} \end{bmatrix} \\
&\overset{\text{def}}{=} \begin{bmatrix} w^{(1,1)}x^{(1)} + w^{(1,2)}x^{(2)} + w^{(1,3)}x^{(3)} \\ w^{(2,1)}x^{(1)} + w^{(2,2)}x^{(2)} + w^{(2,3)}x^{(3)} \end{bmatrix} \\
&= \begin{bmatrix} \mathbf{w}^{(1)}\mathbf{x} \\ \mathbf{w}^{(2)}\mathbf{x} \end{bmatrix}
\end{aligned}
$$

If our matrix had, say, five rows, the result of the product would be a five-dimensional vector.

When the vector is on the left side of the matrix in the multiplication, then it has to be **transposed** before we multiply it by the matrix. The transpose of the vector $\mathbf{x}$ denoted as $\mathbf{x}^{\top}$ makes a row vector out of a column vector. Let's say,

$$
\mathbf{x} = \begin{bmatrix} x^{(1)} \\ x^{(2)} \end{bmatrix}, \text{ then } \mathbf{x}^{\top} \overset{\text{def}}{=} \begin{bmatrix} x^{(1)} & x^{(2)} \end{bmatrix}.
$$

The multiplication of the vector $\mathbf{x}$ by the matrix $\mathbf{W}$ is given by $\mathbf{x}^{\top}\mathbf{W}$,

$$
\begin{aligned}
\mathbf{x}^{\top}\mathbf{W} &= \begin{bmatrix} x^{(1)} & x^{(2)} \end{bmatrix} \begin{bmatrix} w^{(1,1)} & w^{(1,2)} & w^{(1,3)} \\ w^{(2,1)} & w^{(2,2)} & w^{(2,3)} \end{bmatrix} \\
&\overset{\text{def}}{=} \begin{bmatrix} w^{(1,1)}x^{(1)} + w^{(2,1)}x^{(2)}, & w^{(1,2)}x^{(1)} + w^{(2,2)}x^{(2)}, & w^{(1,3)}x^{(1)} + w^{(2,3)}x^{(2)} \end{bmatrix}
\end{aligned}
$$

As you can see, we can only multiply a vector by a matrix if the vector has the same number of dimensions as the number of rows in the matrix.

2.1.6 Functions

A **function** is a relation that associates each element x of a set $\mathcal{X}$, the **domain** of the function, to a single element y of another set $\mathcal{Y}$, the **codomain** of the function. A function usually has a name. If the function is called f, this relation is denoted $y = f(x)$ (read f of x), the element x is the argument or input of the function, and y is the value of the function or the

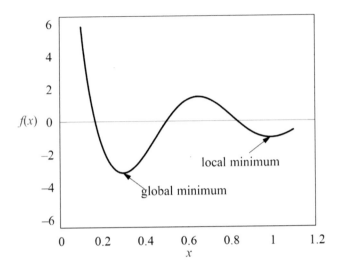

Figure 2.2: A local and a global minima of a function.

output. The symbol that is used for representing the input is the variable of the function (we often say that f is a function of the variable x).

We say that $f(x)$ has a **local minimum** at $x = c$ if $f(x) \geq f(c)$ for every x in some open interval around $x = c$. An **interval** is a set of real numbers with the property that any number that lies between two numbers in the set is also included in the set. An **open interval** does not include its endpoints and is denoted using parentheses. For example, $(0, 1)$ means "all numbers greater than 0 and less than 1". The minimal value among all the local minima is called the **global minimum**. See illustration in Figure 2.2.

A vector function, denoted as $\mathbf{y} = \boldsymbol{f}(x)$ is a function that returns a vector $\mathbf{y}$. It can have a vector or a scalar argument.

2.1.7 Max and Arg Max

Given a set of values $\mathcal{A} = \{a_1, a_2, \ldots, a_n\}$, the operator $\max_{a \in \mathcal{A}} f(a)$ returns the highest value $f(a)$ for all elements in the set $\mathcal{A}$. On the other hand, the operator $\arg \max_{a \in \mathcal{A}} f(a)$ returns the element of the set $\mathcal{A}$ that maximizes $f(a)$.

Sometimes, when the set is implicit or infinite, we can write $\max_a f(a)$ or $\arg \max_a f(a)$.

Operators $\min$ and $\arg \min$ operate in a similar manner.

2.1.8 Assignment Operator

The expression $a \leftarrow f(x)$ means that the variable a gets the new value: the result of $f(x)$. We say that the variable a gets assigned a new value. Similarly, $\mathbf{a} \leftarrow [a_1, a_2]$ means that the vector variable $\mathbf{a}$ gets the two-dimensional vector value $[a_1, a_2]$.

2.1.9 Derivative and Gradient

A **derivative** f' of a function f is a function or a value that describes how fast f grows (or decreases). If the derivative is a constant value, like 5 or -3, then the function grows (or decreases) constantly at any point x of its domain. If the derivative f' is a function, then the function f can grow at a different pace in different regions of its domain. If the derivative f' is positive at some point x, then the function f grows at this point. If the derivative of f is negative at some x, then the function decreases at this point. The derivative of zero at x means that the function's slope at x is horizontal.

The process of finding a derivative is called **differentiation**.

Derivatives for basic functions are known. For example if $f(x) = x^2$, then $f'(x) = 2x$; if $f(x) = 2x$ then $f'(x) = 2$; if $f(x) = 2$ then $f'(x) = 0$ (the derivative of any function $f(x) = c$, where c is a constant value, is zero).

If the function we want to differentiate is not basic, we can find its derivative using the **chain rule**. For instance if $F(x) = f(g(x))$, where f and g are some functions, then $F'(x) = f'(g(x))g'(x)$. For example if $F(x) = (5x + 1)^2$ then $g(x) = 5x + 1$ and $f(g(x)) = (g(x))^2$. By applying the chain rule, we find $F'(x) = 2(5x + 1)g'(x) = 2(5x + 1)5 = 50x + 10$.

Gradient is the generalization of derivative for functions that take several inputs (or one input in the form of a vector or some other complex structure). A gradient of a function is a vector of **partial derivatives**. You can look at finding a partial derivative of a function as the process of finding the derivative by focusing on one of the function's inputs and by considering all other inputs as constant values.

For example, if our function is defined as $f([x^{(1)}, x^{(2)}]) = ax^{(1)} + bx^{(2)} + c$, then the partial derivative of function f *with respect to* $x^{(1)}$, denoted as $\frac{\partial f}{\partial x^{(1)}}$, is given by,

$$\frac{\partial f}{\partial x^{(1)}} = a + 0 + 0 = a,$$

where a is the derivative of the function $ax^{(1)}$; the two zeroes are respectively derivatives of $bx^{(2)}$ and c, because $x^{(2)}$ is considered constant when we compute the derivative with respect to $x^{(1)}$, and the derivative of any constant is zero.

Similarly, the partial derivative of function f with respect to $x^{(2)}$, $\frac{\partial f}{\partial x^{(2)}}$, is given by,

$$\frac{\partial f}{\partial x^{(2)}} = 0 + b + 0 = b.$$

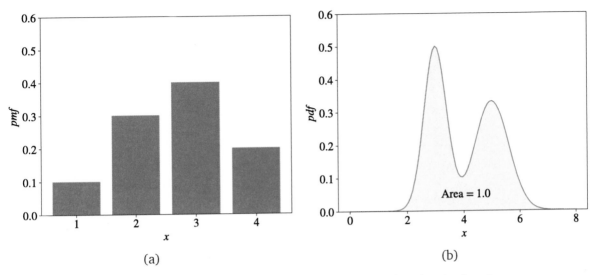

Figure 2.3: A probability mass function and a probability density function.

The gradient of function f, denoted as ∇f is given by the vector $\left[\frac{\partial f}{\partial x^{(1)}}, \frac{\partial f}{\partial x^{(2)}}\right]$.

The chain rule works with partial derivatives too, as I illustrate in Chapter 4.

2.2 Random Variable

A **random variable**, usually written as an italic capital letter, like X, is a variable whose possible values are numerical outcomes of a random phenomenon. Examples of random phenomena with a numerical outcome include a toss of a coin (0 for heads and 1 for tails), a roll of a dice, or the height of the first stranger you meet outside. There are two types of random variables: **discrete** and **continuous**.

A **discrete random variable** takes on only a countable number of distinct values such as *red*, *yellow*, *blue* or 1, 2, 3, ...

The **probability distribution** of a discrete random variable is described by a list of probabilities associated with each of its possible values. This list of probabilities is called a **probability mass function** (pmf). For example: $\Pr(X = red) = 0.3$, $\Pr(X = yellow) = 0.45$, $\Pr(X = blue) = 0.25$. Each probability in a probability mass function is a value greater than or equal to 0. The sum of probabilities equals 1 (Figure 2.3a).

A **continuous random variable** (CRV) takes an infinite number of possible values in some interval. Examples include height, weight, and time. Because the number of values of a continuous random variable X is infinite, the probability $\Pr(X = c)$ for any c is 0. Therefore, instead of the list of probabilities, the probability distribution of a CRV (a continuous probability distribution) is described by a **probability density function** (pdf). The pdf is a function whose codomain is nonnegative and the area under the curve is equal to 1 (Figure 2.3b).

Let a discrete random variable X have k possible values $\{x_i\}_{i=1}^{k}$. The **expectation** of X denoted as $\mathbb{E}[X]$ is given by,

$$\mathbb{E}[X] \overset{\text{def}}{=} \sum_{i=1}^{k} [x_i \cdot \Pr(X = x_i)] \tag{2.1}$$
$$= x_1 \cdot \Pr(X = x_1) + x_2 \cdot \Pr(X = x_2) + \cdots + x_k \cdot \Pr(X = x_k),$$

where $\Pr(X = x_i)$ is the probability that X has the value x_i according to the pmf. The expectation of a random variable is also called the **mean**, **average** or **expected value** and is frequently denoted with the letter μ. The expectation is one of the most important **statistics** of a random variable.

Another important statistic is the **standard deviation**, defined as,

$$\sigma \overset{\text{def}}{=} \sqrt{\mathbb{E}[(X - \mu)^2]}.$$

Variance, denoted as σ^2 or $var(X)$, is defined as,

$$\sigma^2 = \mathbb{E}[(X - \mu)^2].$$

For a discrete random variable, the standard deviation is given by:

$$\sigma = \sqrt{\Pr(X = x_1)(x_1 - \mu)^2 + \Pr(X = x_2)(x_2 - \mu)^2 + \cdots + \Pr(X = x_k)(x_k - \mu)^2},$$

where $\mu = \mathbb{E}[X]$.

The expectation of a continuous random variable X is given by,

$$\mathbb{E}[X] \overset{\text{def}}{=} \int_{\mathbb{R}} x f_X(x) \, dx, \tag{2.2}$$

where f_X is the pdf of the variable X and $\int_{\mathbb{R}}$ is the *integral* of function $x f_X$.

Integral is an equivalent of the summation over all values of the function when the function has a continuous domain. It equals the area under the curve of the function. The property of the pdf that the area under its curve is 1 mathematically means that $\int_{\mathbb{R}} f_X(x) \, dx = 1$.

Most of the time we don't know f_X, but we can observe some values of X. In machine learning, we call these values **examples**, and the collection of these examples is called a **sample** or a **dataset**.

2.3 Unbiased Estimators

Because f_X is usually unknown, but we have a sample $S_X = \{x_i\}_{i=1}^N$, we often content ourselves not with the true values of statistics of the probability distribution, such as expectation, but with their **unbiased estimators**.

We say that $\hat{\theta}(S_X)$ is an unbiased estimator of some statistic θ calculated using a sample S_X drawn from an unknown probability distribution if $\hat{\theta}(S_X)$ has the following property:

$$\mathbb{E}\left[\hat{\theta}(S_X)\right] = \theta,$$

where $\hat{\theta}$ is a **sample statistic**, obtained using a sample S_X and not the real statistic θ that can be obtained only knowing X; the expectation is taken over all possible samples drawn from X. Intuitively, this means that if you can have an unlimited number of such samples as S_X, and you compute some unbiased estimator, such as $\hat{\mu}$, using each sample, then the average of all these $\hat{\mu}$ equals the real statistic μ that you would get computed on X.

It can be shown that an unbiased estimator of an unknown $\mathbb{E}[X]$ (given by either eq. 2.1 or eq. 2.2) is given by $\frac{1}{N}\sum_{i=1}^N x_i$ (called in statistics the **sample mean**).

2.4 Bayes' Rule

The conditional probability $\Pr(X = x|Y = y)$ is the probability of the random variable X to have a specific value x given that another random variable Y has a specific value of y. The **Bayes' Rule** (also known as the **Bayes' Theorem**) stipulates that:

$$\Pr(X = x|Y = y) = \frac{\Pr(Y = y|X = x)\Pr(X = x)}{\Pr(Y = y)}.$$

2.5 Parameter Estimation

Bayes' Rule comes in handy when we have a model of X's distribution, and this model f_θ is a function that has some parameters in the form of a vector θ. An example of such a function could be the Gaussian function that has two parameters, μ and σ, and is defined as:

$$f_\theta(x) = \frac{1}{\sqrt{2\pi\sigma^2}}e^{-\frac{(x-\mu)^2}{2\sigma^2}}, \tag{2.3}$$

where $\theta \stackrel{\text{def}}{=} [\mu, \sigma]$.

This function has all the properties of a pdf[1]. Therefore, we can use it as a model of an unknown distribution of X. We can update the values of parameters in the vector θ from the data using the Bayes' Rule:

$$\Pr(\theta = \hat{\theta}|X = x) \leftarrow \frac{\Pr(X = x|\theta = \hat{\theta})\Pr(\theta = \hat{\theta})}{\Pr(X = x)} = \frac{\Pr(X = x|\theta = \hat{\theta})\Pr(\theta = \hat{\theta})}{\sum_{\tilde{\theta}}\Pr(X = x|\theta = \tilde{\theta})\Pr(\theta = \tilde{\theta})}. \quad (2.4)$$

where $\Pr(X = x|\theta = \hat{\theta}) \stackrel{\text{def}}{=} f_{\hat{\theta}}$.

If we have a sample $\mathcal{S}$ of X and the set of possible values for θ is finite, we can easily estimate $\Pr(\theta = \hat{\theta})$ by applying Bayes' Rule iteratively, one example $x \in \mathcal{S}$ at a time. The initial value $\Pr(\theta = \hat{\theta})$ can be guessed such that $\sum_{\hat{\theta}}\Pr(\theta = \hat{\theta}) = 1$. This guess of the probabilities for different $\hat{\theta}$ is called the **prior**.

First, we compute $\Pr(\theta = \hat{\theta}|X = x_1)$ for all possible values $\hat{\theta}$. Then, before updating $\Pr(\theta = \hat{\theta}|X = x)$ once again, this time for $x = x_2 \in \mathcal{S}$ using eq. 2.4, we replace the prior $\Pr(\theta = \hat{\theta})$ in eq. 2.4 by the new estimate $\Pr(\theta = \hat{\theta}) \leftarrow \frac{1}{N}\sum_{x \in \mathcal{S}}\Pr(\theta = \hat{\theta}|X = x)$.

The best value of the parameters θ^* given one example is obtained using the principle of **maximum a posteriori** (or MAP):

$$\theta^* = \arg\max_{\theta} \prod_{i=1}^{N} \Pr(\theta = \hat{\theta}|X = x_i). \quad (2.5)$$

If the set of possible values for θ isn't finite, then we need to optimize eq. 2.5 directly using a numerical optimization routine, such as gradient descent, which we consider in Chapter 4. Usually, we optimize the natural logarithm of the right-hand side expression in eq. 2.5 because the logarithm of a product becomes the sum of logarithms and it's easier for the machine to work with a sum than with a product[2].

2.6   Parameters vs. Hyperparameters

A hyperparameter is a property of a learning algorithm, usually (but not always) having a numerical value. That value influences the way the algorithm works. Hyperparameters aren't learned by the algorithm itself from data. They have to be set by the data analyst before running the algorithm. I show how to do that in Chapter 5.

---

[1]In fact, eq. 2.3 defines the pdf of one of the most frequently used in practice probability distributions called **Gaussian distribution** or **normal distribution** and denoted as $\mathcal{N}(\mu, \sigma^2)$.

[2]Multiplication of many numbers can give either a very small result or a very large one. It often results in the problem of numerical overflow when the machine cannot store such extreme numbers in memory.

Parameters are variables that define the model learned by the learning algorithm. Parameters are directly modified by the learning algorithm based on the training data. The goal of learning is to find such values of parameters that make the model optimal in a certain sense.

2.7 Classification vs. Regression

Classification is a problem of automatically assigning a **label** to an **unlabeled example**. Spam detection is a famous example of classification.

In machine learning, the classification problem is solved by a **classification learning algorithm** that takes a collection of **labeled examples** as inputs and produces a **model** that can take an unlabeled example as input and either directly output a label or output a number that can be used by the analyst to deduce the label. An example of such a number is a probability.

In a classification problem, a label is a member of a finite set of **classes**. If the size of the set of classes is two ("sick"/"healthy", "spam"/"not_spam"), we talk about **binary classification** (also called **binomial** in some sources). **Multiclass classification** (also called **multinomial**) is a classification problem with three or more classes[3].

While some learning algorithms naturally allow for more than two classes, others are by nature binary classification algorithms. There are strategies allowing to turn a binary classification learning algorithm into a multiclass one. I talk about one of them in Chapter 7.

Regression is a problem of predicting a real-valued label (often called a **target**) given an unlabeled example. Estimating house price valuation based on house features, such as area, the number of bedrooms, location and so on is a famous example of regression.

The regression problem is solved by a **regression learning algorithm** that takes a collection of labeled examples as inputs and produces a model that can take an unlabeled example as input and output a target.

2.8 Model-Based vs. Instance-Based Learning

Most supervised learning algorithms are model-based. We have already seen one such algorithm: SVM. Model-based learning algorithms use the training data to create a **model** that has **parameters** learned from the training data. In SVM, the two parameters we saw were w^* and b^*. After the model was built, the training data can be discarded.

Instance-based learning algorithms use the whole dataset as the model. One instance-based algorithm frequently used in practice is **k-Nearest Neighbors** (kNN). In classification, to predict a label for an input example the kNN algorithm looks at the close neighborhood of the input example in the space of feature vectors and outputs the label that it saw the most often in this close neighborhood.

---

[3]There's still one label per example though.

2.9 Shallow vs. Deep Learning

A **shallow learning** algorithm learns the parameters of the model directly from the features of the training examples. Most supervised learning algorithms are shallow. The notorious exceptions are **neural network** learning algorithms, specifically those that build neural networks with more than one **layer** between input and output. Such neural networks are called **deep neural networks**. In deep neural network learning (or, simply, **deep learning**), contrary to shallow learning, most model parameters are learned not directly from the features of the training examples, but from the outputs of the preceding layers.

Don't worry if you don't understand what that means right now. We look at neural networks more closely in Chapter 6.

Chapter 3

Fundamental Algorithms

In this chapter, I describe five algorithms which are not just the most known but also either very effective on their own or are used as building blocks for the most effective learning algorithms out there.

3.1 Linear Regression

Linear regression is a popular regression learning algorithm that learns a model which is a linear combination of features of the input example.

3.1.1 Problem Statement

We have a collection of labeled examples $\{(\mathbf{x}_i, y_i)\}_{i=1}^N$, where N is the size of the collection, $\mathbf{x}_i$ is the D-dimensional feature vector of example $i = 1, \ldots, N$, y_i is a real-valued[1] target and every feature $x_i^{(j)}$, $j = 1, \ldots, D$, is also a real number.

We want to build a model $f_{\mathbf{w},b}(\mathbf{x})$ as a linear combination of features of example $\mathbf{x}$:

$$f_{\mathbf{w},b}(\mathbf{x}) = \mathbf{w}\mathbf{x} + b, \tag{3.1}$$

where $\mathbf{w}$ is a D-dimensional vector of parameters and b is a real number. The notation $f_{\mathbf{w},b}$ means that the model f is parametrized by two values: $\mathbf{w}$ and b.

We will use the model to predict the unknown y for a given $\mathbf{x}$ like this: $y \leftarrow f_{\mathbf{w},b}(\mathbf{x})$. Two models parametrized by two different pairs $(\mathbf{w}, b)$ will likely produce two different predictions

---

[1]To say that y_i is real-valued, we write $y_i \in \mathbb{R}$, where $\mathbb{R}$ denotes the set of all real numbers, an infinite set of numbers from minus infinity to plus infinity.

when applied to the same example. We want to find the optimal values $(\mathbf{w}^*, b^*)$. Obviously, the optimal values of parameters define the model that makes the most accurate predictions.

You could have noticed that the form of our linear model in eq. 3.1 is very similar to the form of the SVM model. The only difference is the missing sign operator. The two models are indeed similar. However, the hyperplane in the SVM plays the role of the decision boundary: it's used to separate two groups of examples from one another. As such, it has to be as far from each group as possible.

On the other hand, the hyperplane in linear regression is chosen to be as close to all training examples as possible.

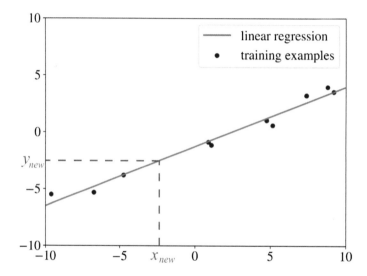

Figure 3.1: Linear Regression for one-dimensional examples.

You can see why this latter requirement is essential by looking at the illustration in Figure 3.1. It displays the regression line (in red) for one-dimensional examples (blue dots). We can use this line to predict the value of the target y_{new} for a new unlabeled input example x_{new}. If our examples are D-dimensional feature vectors (for $D > 1$), the only difference with the one-dimensional case is that the regression model is not a line but a plane (for two dimensions) or a hyperplane (for $D > 2$).

Now you see why it's essential to have the requirement that the regression hyperplane lies as close to the training examples as possible: if the red line in Figure 3.1 was far from the blue dots, the prediction y_{new} would have fewer chances to be correct.

3.1.2 Solution

To get this latter requirement satisfied, the optimization procedure which we use to find the optimal values for $\mathbf{w}^*$ and b^* tries to minimize the following expression:

$$\frac{1}{N} \sum_{i=1...N} (f_{\mathbf{w},b}(\mathbf{x}_i) - y_i)^2. \tag{3.2}$$

In mathematics, the expression we minimize or maximize is called an **objective function**, or, simply, an **objective**. The expression $(f_{\mathbf{w},b}(\mathbf{x}_i) - y_i)^2$ in the above objective is called the **loss function**. It's a measure of penalty for misclassification of example i. This particular choice of the loss function is called **squared error loss**. All model-based learning algorithms have a loss function and what we do to find the best model is we try to minimize the objective known as the **cost function**. In linear regression, the cost function is given by the average loss, also called the **empirical risk**. The average loss, or empirical risk, for a model, is the average of all penalties obtained by applying the model to the training data.

Why is the loss in linear regression a quadratic function? Why couldn't we get the absolute value of the difference between the true target y_i and the predicted value $f(\mathbf{x}_i)$ and use that as a penalty? We could. Moreover, we also could use a cube instead of a square.

Now you probably start realizing how many seemingly arbitrary decisions are made when we design a machine learning algorithm: we decided to use the linear combination of features to predict the target. However, we could use a square or some other polynomial to combine the values of features. We could also use some other loss function that makes sense: the absolute difference between $f(\mathbf{x}_i)$ and y_i makes sense, the cube of the difference too; the **binary loss** (1 when $f(\mathbf{x}_i)$ and y_i are different and 0 when they are the same) also makes sense, right?

If we made different decisions about the form of the model, the form of the loss function, and about the choice of the algorithm that minimizes the average loss to find the best values of parameters, we would end up inventing a different machine learning algorithm. Sounds easy, doesn't it? However, do not rush to invent a new learning algorithm. The fact that it's different doesn't mean that it will work better in practice.

People invent new learning algorithms for one of the two main reasons:

1. The new algorithm solves a specific practical problem better than the existing algorithms.
2. The new algorithm has better theoretical guarantees on the quality of the model it produces.

One practical justification of the choice of the linear form for the model is that it's simple. Why use a complex model when you can use a simple one? Another consideration is that linear models rarely overfit. **Overfitting** is the property of a model such that the model predicts very well labels of the examples used during training but frequently makes errors when applied to examples that weren't seen by the learning algorithm during training.

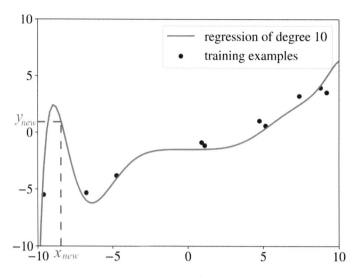

Figure 3.2: Overfitting.

An example of overfitting in regression is shown in Figure 3.2. The data used to build the red regression line is the same as in Figure 3.1. The difference is that this time, this is the polynomial regression with a polynomial of degree 10. The regression line predicts almost perfectly the targets almost all training examples, but will likely make significant errors on new data, as you can see in Figure 3.1 for x_{new}. We talk more about overfitting and how to avoid it in Chapter 5.

Now you know why linear regression can be useful: it doesn't overfit much. But what about the squared loss? Why did we decide that it should be squared? In 1705, the French mathematician Adrien-Marie Legendre, who first published the sum of squares method for gauging the quality of the model stated that squaring the error before summing is *convenient*. Why did he say that? The absolute value is not convenient, because it doesn't have a continuous derivative, which makes the function not smooth. Functions that are not smooth create unnecessary difficulties when employing linear algebra to find closed form solutions to optimization problems. Closed form solutions to finding an optimum of a function are simple algebraic expressions and are often preferable to using complex numerical optimization methods, such as **gradient descent** (used, among others, to train neural networks).

Intuitively, squared penalties are also advantageous because they exaggerate the difference between the true target and the predicted one according to the value of this difference. We might also use the powers 3 or 4, but their derivatives are more complicated to work with.

Finally, why do we care about the derivative of the average loss? If we can calculate the gradient of the function in eq. 3.2, we can then set this gradient to zero[2] and find the solution

---

[2]To find the minimum or the maximum of a function, we set the gradient to zero because the value of the gradient at extrema of a function is always zero. In 2D, the gradient at an extremum is a horizontal line.

to a system of equations that gives us the optimal values $\mathbf{w}^*$ and b^*.

3.2 Logistic Regression

The first thing to say is that logistic regression is not a regression, but a classification learning algorithm. The name comes from statistics and is due to the fact that the mathematical formulation of logistic regression is similar to that of linear regression.

I explain logistic regression on the case of binary classification. However, it can naturally be extended to multiclass classification.

3.2.1 Problem Statement

In **logistic regression**, we still want to model y_i as a linear function of $\mathbf{x}_i$, however, with a binary y_i this is not straightforward. The linear combination of features such as $\mathbf{w}\mathbf{x}_i + b$ is a function that spans from minus infinity to plus infinity, while y_i has only two possible values.

At the time where the absence of computers required scientists to perform manual calculations, they were eager to find a linear classification model. They figured out that if we define a negative label as 0 and the positive label as 1, we would just need to find a simple continuous function whose codomain is $(0, 1)$. In such a case, if the value returned by the model for input $\mathbf{x}$ is closer to 0, then we assign a negative label to $\mathbf{x}$; otherwise, the example is labeled as positive. One function that has such a property is the **standard logistic function** (also known as the **sigmoid function**):

$$f(x) = \frac{1}{1 + e^{-x}},$$

where e is the base of the natural logarithm (also called *Euler's number*; e^x is also known as the *exp(x)* function in programming languages). Its graph is depicted in Figure 3.3.

The logistic regression model looks like this:

$$f_{\mathbf{w},b}(\mathbf{x}) \stackrel{\text{def}}{=} \frac{1}{1 + e^{-(\mathbf{w}\mathbf{x}+b)}}. \tag{3.3}$$

You can see the familiar term $\mathbf{w}\mathbf{x} + b$ from linear regression.

By looking at the graph of the standard logistic function, we can see how well it fits our classification purpose: if we optimize the values of $\mathbf{w}$ and b appropriately, we could interpret the output of $f(\mathbf{x})$ as the probability of y_i being positive. For example, if it's higher than or equal to the threshold 0.5 we would say that the class of $\mathbf{x}$ is positive; otherwise, it's negative. In practice, the choice of the threshold could be different depending on the problem. We return to this discussion in Chapter 5 when we talk about model performance assessment.

Now, how do we find optimal $\mathbf{w}^*$ and b^*? In linear regression, we minimized the empirical risk which was defined as the average squared error loss, also known as the **mean squared error** or MSE.

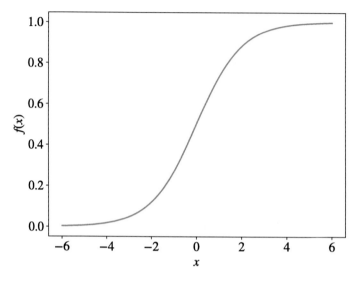

Figure 3.3: Standard logistic function.

3.2.2 Solution

In logistic regression, on the other hand, we maximize the **likelihood** of our training set according to the model. In statistics, the likelihood function defines how likely the observation (an example) is according to our model.

For instance, let's have a labeled example $(\mathbf{x}_i, y_i)$ in our training data. Assume also that we found (guessed) some specific values $\hat{\mathbf{w}}$ and $\hat{b}$ of our parameters. If we now apply our model $f_{\hat{\mathbf{w}}, \hat{b}}$ to $\mathbf{x}_i$ using eq. 3.3 we will get some value $0 < p < 1$ as output. If y_i is the positive class, the likelihood of y_i being the positive class, according to our model, is given by p. Similarly, if y_i is the negative class, the likelihood of it being the negative class is given by $1 - p$.

The optimization criterion in logistic regression is called **maximum likelihood**. Instead of minimizing the average loss, like in linear regression, we now maximize the likelihood of the training data according to our model:

$$L_{\mathbf{w},b} \overset{\text{def}}{=} \prod_{i=1...N} f_{\mathbf{w},b}(\mathbf{x}_i)^{y_i} (1 - f_{\mathbf{w},b}(\mathbf{x}_i))^{(1-y_i)}. \tag{3.4}$$

The expression $f_{\mathbf{w},b}(\mathbf{x})^{y_i} (1 - f_{\mathbf{w},b}(\mathbf{x}))^{(1-y_i)}$ may look scary but it's just a fancy mathematical way of saying: "$f_{\mathbf{w},b}(\mathbf{x})$ when $y_i = 1$ and $(1 - f_{\mathbf{w},b}(\mathbf{x}))$ otherwise". Indeed, if $y_i = 1$, then

$(1 - f_{\mathbf{w},b}(\mathbf{x}))^{(1-y_i)}$ equals 1 because $(1 - y_i) = 0$ and we know that anything power 0 equals 1. On the other hand, if $y_i = 0$, then $f_{\mathbf{w},b}(\mathbf{x})^{y_i}$ equals 1 for the same reason.

You may have noticed that we used the product operator $\prod$ in the objective function instead of the sum operator $\sum$ which was used in linear regression. It's because the likelihood of observing N labels for N examples is the product of likelihoods of each observation (assuming that all observations are independent of one another, which is the case). You can draw a parallel with the multiplication of probabilities of outcomes in a series of independent experiments in the probability theory.

Because of the exp function used in the model, in practice, it's more convenient to maximize the **log-likelihood** instead of likelihood. The log-likelihood is defined as follows:

$$LogL_{\mathbf{w},b} \overset{\text{def}}{=} \ln(L_{\mathbf{w},b}(\mathbf{x})) = \sum_{i=1}^{N} [y_i \ln f_{\mathbf{w},b}(\mathbf{x}) + (1 - y_i) \ln (1 - f_{\mathbf{w},b}(\mathbf{x}))].$$

Because ln is a **strictly increasing function**, maximizing this function is the same as maximizing its argument, and the solution to this new optimization problem is the same as the solution to the original problem.

Contrary to linear regression, there's no closed form solution to the above optimization problem. A typical numerical optimization procedure used in such cases is **gradient descent**. We talk about it in the next chapter.

3.3   Decision Tree Learning

A **decision tree** is an acyclic **graph** that can be used to make decisions. In each branching node of the graph, a specific feature j of the feature vector is examined. If the value of the feature is below a specific threshold, then the left branch is followed; otherwise, the right branch is followed. As the leaf node is reached, the decision is made about the class to which the example belongs.

As the title of the section suggests, a decision tree can be learned from data.

3.3.1   Problem Statement

Like previously, we have a collection of labeled examples; labels belong to the set $\{0, 1\}$. We want to build a decision tree that would allow us to predict the class given a feature vector.

3.3.2   Solution

There are various formulations of the decision tree learning algorithm. In this book, we consider just one, called **ID3**.

The optimization criterion, in this case, is the average log-likelihood:

$$\frac{1}{N}\sum_{i=1}^{N}\left[y_i \ln f_{ID3}(\mathbf{x}_i) + (1 - y_i)\ln\left(1 - f_{ID3}(\mathbf{x}_i)\right)\right],\qquad(3.5)$$

where f_{ID3} is a decision tree.

By now, it looks very similar to logistic regression. However, contrary to the logistic regression learning algorithm which builds a **parametric model** $f_{\mathbf{w}^*,b^*}$ by finding an *optimal solution* to the optimization criterion, the ID3 algorithm optimizes it *approximately* by constructing a **nonparametric model** $f_{ID3}(\mathbf{x}) \stackrel{\text{def}}{=} \Pr(y = 1|\mathbf{x})$.

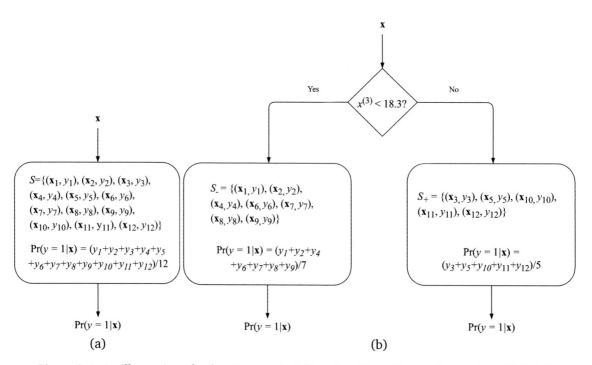

Figure 3.4: An illustration of a decision tree building algorithm. The set $\mathcal{S}$ contains 12 labeled examples. (a) In the beginning, the decision tree only contains the start node; it makes the same prediction for any input. (b) The decision tree after the first split; it tests whether feature 3 is less than 18.3 and, depending on the result, the prediction is made in one of the two leaf nodes.

The ID3 learning algorithm works as follows. Let $\mathcal{S}$ denote a set of labeled examples. In the beginning, the decision tree only has a start node that contains all examples: $\mathcal{S} \stackrel{\text{def}}{=} \{(\mathbf{x}_i, y_i)\}_{i=1}^{N}$. Start with a constant model $f_{ID3}^{\mathcal{S}}$ defined as,

$$f_{ID3}^{\mathcal{S}} \stackrel{\text{def}}{=} \frac{1}{|\mathcal{S}|} \sum_{(\mathbf{x},y) \in \mathcal{S}} y. \tag{3.6}$$

The prediction given by the above model, $f_{ID3}^{\mathcal{S}}(\mathbf{x})$, would be the same for any input $\mathbf{x}$. The corresponding decision tree built using a toy dataset of 12 labeled examples is shown in Figure 3.4a.

Then we search through all features $j = 1, \ldots, D$ and all thresholds t, and split the set S into two subsets: $\mathcal{S}_- \stackrel{\text{def}}{=} \{(\mathbf{x},y) \mid (\mathbf{x},y) \in S, x^{(j)} < t\}$ and $\mathcal{S}_+ \stackrel{\text{def}}{=} \{(\mathbf{x},y) \mid (\mathbf{x},y) \in S, x^{(j)} \geq t\}$. The two new subsets would go to two new leaf nodes, and we evaluate, for all possible pairs (j, t) how good the split with pieces $\mathcal{S}_-$ and $\mathcal{S}_+$ is. Finally, we pick the best such values (j, t), split S into $\mathcal{S}_+$ and $\mathcal{S}_-$, form two new leaf nodes, and continue recursively on $\mathcal{S}_+$ and $\mathcal{S}_-$ (or quit if no split produces a model that's sufficiently better than the current one). A decision tree after one split is illustrated in Figure 3.4b.

Now you should wonder what do the words "evaluate how good the split is" mean. In ID3, the goodness of a split is estimated by using the criterion called **entropy**. Entropy is a measure of uncertainty about a random variable. It reaches its maximum when all values of the random variables are equiprobable. Entropy reaches its minimum when the random variable can have only one value. The entropy of a set of examples S is given by,

$$H(\mathcal{S}) \stackrel{\text{def}}{=} -f_{ID3}^{\mathcal{S}} \ln f_{ID3}^{\mathcal{S}} - (1 - f_{ID3}^{\mathcal{S}}) \ln(1 - f_{ID3}^{\mathcal{S}}).$$

When we split a set of examples by a certain feature j and a threshold t, the entropy of a split, $H(\mathcal{S}_-, \mathcal{S}_+)$, is simply a weighted sum of two entropies:

$$H(\mathcal{S}_-, \mathcal{S}_+) \stackrel{\text{def}}{=} \frac{|\mathcal{S}_-|}{|\mathcal{S}|} H(\mathcal{S}_-) + \frac{|\mathcal{S}_+|}{|\mathcal{S}|} H(\mathcal{S}_+). \tag{3.7}$$

So, in ID3, at each step, at each leaf node, we find a split that minimizes the entropy given by eq. 3.7 or we stop at this leaf node.

The algorithm stops at a leaf node in any of the below situations:

- All examples in the leaf node are classified correctly by the one-piece model (eq. 3.6).
- We cannot find an attribute to split upon.
- The split reduces the entropy less than some ϵ (the value for which has to be found experimentally[3]).
- The tree reaches some maximum depth d (also has to be found experimentally).

---

[3]In Chapter 5, I show how to do that in the section on hyperparameter tuning.

Because in ID3, the decision to split the dataset on each iteration is local (doesn't depend on future splits), the algorithm doesn't guarantee an optimal solution. The model can be improved by using techniques like *backtracking* during the search for the optimal decision tree at the cost of possibly taking longer to build a model.

The most widely used formulation of a decision tree learning algorithm is called **C4.5**. It has several additional features as compared to ID3:

- it accepts both continuous and discrete features;
- it handles incomplete examples;
- it solves overfitting problem by using a bottom-up technique known as "pruning".

Pruning consists of going back through the tree once it's been created and removing branches that don't contribute significantly enough to the error reduction by replacing them with leaf nodes.

The entropy-based split criterion intuitively makes sense: entropy reaches its minimum of 0 when all examples in S have the same label; on the other hand, the entropy is at its maximum of 1 when exactly one-half of examples in S is labeled with 1, making such a leaf useless for classification. The only remaining question is how this algorithm approximately maximizes the average log-likelihood criterion. I leave it for further reading.

3.4 Support Vector Machine

I already presented SVM in the introduction, so this section only fills a couple of blanks. Two critical questions need to be answered:

1. What if there's noise in the data and no hyperplane can perfectly separate positive examples from negative ones?
2. What if the data cannot be separated using a plane, but could be separated by a higher-order polynomial?

You can see both situations depicted in Figure 3.5. In the left case, the data could be separated by a straight line if not for the noise (outliers or examples with wrong labels). In the right case, the decision boundary is a circle and not a straight line.

Remember that in SVM, we want to satisfy the following constraints:

$$\mathbf{w}x_i - b \geq +1 \quad \text{if } y_i = +1,$$
$$\mathbf{w}x_i - b \leq -1 \quad \text{if } y_i = -1. \tag{3.8}$$

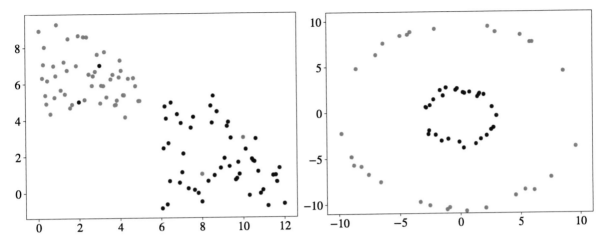

Figure 3.5: Linearly non-separable cases. Left: the presence of noise. Right: inherent nonlinearity.

We also want to minimize $\|\mathbf{w}\|$ so that the hyperplane is equally distant from the closest examples of each class. Minimizing $\|\mathbf{w}\|$ is equivalent to minimizing $\frac{1}{2}\|\mathbf{w}\|^2$, and the use of this term makes it possible to perform quadratic programming optimization later on. The optimization problem for SVM, therefore, looks like this:

$$\min \frac{1}{2}\|\mathbf{w}\|^2, \text{ such that } y_i(\mathbf{x}_i\mathbf{w} - b) - 1 \geq 0, i = 1, \ldots, N. \tag{3.9}$$

3.4.1 Dealing with Noise

To extend SVM to cases in which the data is not linearly separable, we introduce the **hinge loss** function: $\max(0, 1 - y_i(\mathbf{w}\mathbf{x}_i - b))$.

The hinge loss function is zero if the constraints in 3.8 are satisfied; in other words, if $\mathbf{w}\mathbf{x}_i$ lies on the correct side of the decision boundary. For data on the wrong side of the decision boundary, the function's value is proportional to the distance from the decision boundary.

We then wish to minimize the following cost function,

$$C\|\mathbf{w}\|^2 + \frac{1}{N}\sum_{i=1}^{N}\max(0, 1 - y_i(\mathbf{w}\mathbf{x}_i - b)),$$

where the hyperparameter C determines the tradeoff between increasing the size of the decision boundary and ensuring that each $\mathbf{x}_i$ lies on the correct side of the decision boundary. The value of C is usually chosen experimentally, just like ID3's hyperparameters ϵ and d.

SVMs that optimize hinge loss are called *soft-margin* SVMs, while the original formulation is referred to as a *hard-margin* SVM.

As you can see, for sufficiently high values of C, the second term in the cost function will become negligible, so the SVM algorithm will try to find the highest margin by completely ignoring misclassification. As we decrease the value of C, making classification errors is becoming more costly, so the SVM algorithm tries to make fewer mistakes by sacrificing the margin size. As we have already discussed, a larger margin is better for generalization. Therefore, C regulates the tradeoff between classifying the training data well (minimizing empirical risk) and classifying future examples well (generalization).

3.4.2   Dealing with Inherent Non-Linearity

SVM can be adapted to work with datasets that cannot be separated by a hyperplane in its original space. Indeed, if we manage to transform the original space into a space of higher dimensionality, we could hope that the examples will become linearly separable in this transformed space. In SVMs, using a function to *implicitly* transform the original space into a higher dimensional space during the cost function optimization is called the **kernel trick.**

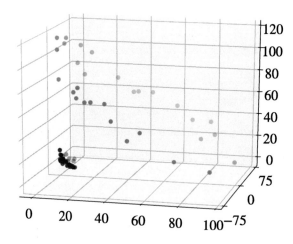

Figure 3.6: The data from Figure 3.5 (right) becomes linearly separable after a transformation into a three-dimensional space.

The effect of applying the kernel trick is illustrated in Figure 3.6. As you can see, it's possible to transform a two-dimensional non-linearly-separable data into a linearly-separable three-dimensional data using a specific mapping $\phi : \mathbf{x} \mapsto \phi(\mathbf{x})$, where $\phi(\mathbf{x})$ is a vector of higher dimensionality than **x**. For the example of 2D data in Figure 3.5 (right), the mapping ϕ for that projects a 2D example $\mathbf{x} = [q, p]$ into a 3D space (Figure 3.6) would look like this: $\phi([q, p]) \stackrel{\text{def}}{=} (q^2, \sqrt{2}qp, p^2)$, where $\cdot^2$ means $\cdot$ squared. You see now that the data becomes linearly separable in the transformed space.

However, we don't know a priori which mapping ϕ would work for our data. If we first transform all our input examples using some mapping into very high dimensional vectors and then apply SVM to this data, and we try all possible mapping functions, the computation could become very inefficient, and we would never solve our classification problem.

Fortunately, scientists figured out how to use **kernel functions** (or, simply, **kernels**) to efficiently work in higher-dimensional spaces *without doing this transformation explicitly*. To understand how kernels work, we have to see first how the optimization algorithm for SVM finds the optimal values for **w** and b.

The method traditionally used to solve the optimization problem in eq. 3.9 is the *method of Lagrange multipliers*. Instead of solving the original problem from eq. 3.9, it is convenient to solve an equivalent problem formulated like this:

$$\max_{\alpha_1 \dots \alpha_N} \sum_{i=1}^{N} \alpha_i - \frac{1}{2} \sum_{i=1}^{N} \sum_{k=1}^{N} y_i \alpha_i (\mathbf{x}_i \mathbf{x}_k) y_k \alpha_k \text{ subject to } \sum_{i=1}^{N} \alpha_i y_i = 0 \text{ and } \alpha_i \geq 0, i = 1, \dots, N,$$

where α_i are called Lagrange multipliers. When formulated like this, the optimization problem becomes a convex quadratic optimization problem, efficiently solvable by quadratic programming algorithms.

Now, you could have noticed that in the above formulation, there is a term $\mathbf{x}_i \mathbf{x}_k$, and this is the only place where the feature vectors are used. If we want to transform our vector space into higher dimensional space, we need to transform $\mathbf{x}_i$ into $\phi(\mathbf{x}_i)$ and $\mathbf{x}_k$ into $\phi(\mathbf{x}_k)$ and then multiply $\phi(\mathbf{x}_i)$ and $\phi(\mathbf{x}_k)$. Doing so would be very costly.

On the other hand, we are only interested in the result of the dot-product $\mathbf{x}_i \mathbf{x}_k$, which, as we know, is a real number. We don't care how this number was obtained as long as it's correct. By using the kernel trick, we can get rid of a costly transformation of original feature vectors into higher-dimensional vectors and avoid computing their dot-product. We replace that by a simple operation on the original feature vectors that gives the same result. For example, instead of transforming (q_1, p_1) into $(q_1^2, \sqrt{2}q_1 p_1, p_1^2)$ and (q_2, p_2) into $(q_2^2, \sqrt{2}q_2 p_2, p_2^2)$ and then computing the dot-product of $(q_1^2, \sqrt{2}q_1 p_1, p_1^2)$ and $(q_2^2, \sqrt{2}q_2 p_2, p_2^2)$ to obtain $(q_1^2 q_2^2 + 2q_1 q_2 p_1 p_2 + p_1^2 p_2^2)$ we could find the dot-product between (q_1, p_1) and (q_2, p_2) to get $(q_1 q_2 + p_1 p_2)$ and then square it to get exactly the same result $(q_1^2 q_2^2 + 2q_1 q_2 p_1 p_2 + p_1^2 p_2^2)$.

That was an example of the kernel trick, and we used the quadratic kernel $k(\mathbf{x}_i, \mathbf{x}_k) \stackrel{\text{def}}{=} (\mathbf{x}_i \mathbf{x}_k)^2$. Multiple kernel functions exist, the most widely used of which is the **RBF kernel**:

$$k(\mathbf{x}, \mathbf{x}') = \exp\left(-\frac{\|\mathbf{x} - \mathbf{x}'\|^2}{2\sigma^2}\right),$$

where $\|\mathbf{x} - \mathbf{x}'\|^2$ is the squared **Euclidean distance** between two feature vectors. The Euclidean distance is given by the following equation:

$$d(\mathbf{x}_i, \mathbf{x}_k) \overset{\text{def}}{=} \sqrt{\left(x_i^{(1)} - x_k^{(1)}\right)^2 + \left(x_i^{(2)} - x_k^{(2)}\right)^2 + \cdots + \left(x_i^{(N)} - x_k^{(N)}\right)^2} = \sqrt{\sum_{j=1}^{D} \left(x_i^{(j)} - x_k^{(j)}\right)^2}.$$

It can be shown that the feature space of the RBF (for "radial basis function") kernel has an infinite number of dimensions. By varying the hyperparameter σ, the data analyst can choose between getting a smooth or curvy decision boundary in the original space.

3.5   k-Nearest Neighbors

k-Nearest Neighbors (kNN) is a non-parametric learning algorithm. Contrary to other learning algorithms that allow discarding the training data after the model is built, kNN keeps all training examples in memory. Once a new, previously unseen example $\mathbf{x}$ comes in, the kNN algorithm finds k training examples closest to $\mathbf{x}$ and returns the majority label, in case of classification, or the average label, in case of regression.

The closeness of two examples is given by a distance function. For example, Euclidean distance seen above is frequently used in practice. Another popular choice of the distance function is the negative **cosine similarity**. Cosine similarity defined as,

$$s(\mathbf{x}_i, \mathbf{x}_k) \overset{\text{def}}{=} \cos(\angle(\mathbf{x}_i, \mathbf{x}_k)) = \frac{\sum_{j=1}^{D} x_i^{(j)} x_k^{(j)}}{\sqrt{\sum_{j=1}^{D} \left(x_i^{(j)}\right)^2} \sqrt{\sum_{j=1}^{D} \left(x_k^{(j)}\right)^2}},$$

is a measure of similarity of the directions of two vectors. If the angle between two vectors is 0 degrees, then two vectors point to the same direction, and cosine similarity is equal to 1. If the vectors are orthogonal, the cosine similarity is 0. For vectors pointing in opposite directions, the cosine similarity is -1. If we want to use cosine similarity as a distance metric, we need to multiply it by -1. Other popular distance metrics include Chebychev distance, Mahalanobis distance, and Hamming distance. The choice of the distance metric, as well as the value for k, are the choices the analyst makes before running the algorithm. So these are hyperparameters. The distance metric could also be learned from data (as opposed to guessing it). We talk about that in Chapter 10.

Chapter 4

Anatomy of a Learning Algorithm

4.1 Building Blocks of a Learning Algorithm

You may have noticed by reading the previous chapter that each learning algorithm we saw consisted of three parts:

1) a loss function;
2) an optimization criterion based on the loss function (a cost function, for example); and
3) an optimization routine leveraging training data to find a solution to the optimization criterion.

These are the building blocks of any learning algorithm. You saw in the previous chapter that some algorithms were designed to explicitly optimize a specific criterion (both linear and logistic regressions, SVM). Some others, including decision tree learning and kNN, optimize the criterion implicitly. Decision tree learning and kNN are among the oldest machine learning algorithms and were invented experimentally based on intuition, without a specific global optimization criterion in mind, and (like it often happened in scientific history) the optimization criteria were developed later to explain why those algorithms work.

By reading modern literature on machine learning, you often encounter references to **gradient descent** or **stochastic gradient descent**. These are two most frequently used optimization algorithms used in cases where the optimization criterion is differentiable.

Gradient descent is an iterative optimization algorithm for finding the minimum of a function. To find a *local* minimum of a function using gradient descent, one starts at some random point and takes steps proportional to the negative of the gradient (or approximate gradient) of the function at the current point.

Gradient descent can be used to find optimal parameters for linear and logistic regression, SVM and also neural networks which we consider later. For many models, such as logistic regression or SVM, the optimization criterion is *convex*. Convex functions have only one minimum, which is global. Optimization criteria for neural networks are not convex, but in practice even finding a local minimum suffices.

Let's see how gradient descent works.

4.2   Gradient Descent

In this section, I demonstrate how gradient descent finds the solution to a linear regression problem[1]. I illustrate my description with Python code as well as with plots that show how the solution improves after some iterations of gradient descent. I use a dataset with only one feature. However, the optimization criterion will have two parameters: w and b. The extension to multi-dimensional training data is straightforward: you have variables $w^{(1)}$, $w^{(2)}$, and b for two-dimensional data, $w^{(1)}$, $w^{(2)}$, $w^{(3)}$, and b for three-dimensional data and so on.

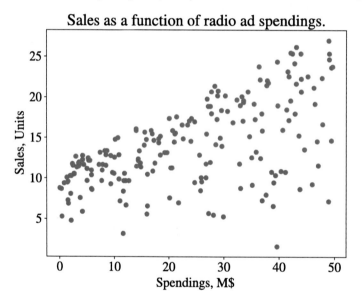

Figure 4.1: The original data. The Y-axis corresponds to the sales in units (the quantity we want to predict), the X-axis corresponds to our feature: the spendings on radio ads in M\$.

To give a practical example, I use the real dataset (can be found on the book's wiki) with the following columns: the Spendings of various companies on radio advertising each year and

---

[1]As you know, linear regression has a closed form solution. That means that gradient descent is not needed to solve this specific type of problem. However, for illustration purposes, linear regression is a perfect problem to explain gradient descent.

their annual Sales in terms of units sold. We want to build a regression model that we can use to predict units sold based on how much a company spends on radio advertising. Each row in the dataset represents one specific company:

Company	Spendings, M$	Sales, Units
1	37.8	22.1
2	39.3	10.4
3	45.9	9.3
4	41.3	18.5
..	..	..

We have data for 200 companies, so we have 200 training examples in the form $(x_i, y_i) = (Spendings_i, Sales_i)$. Figure 4.1 shows all examples on a 2D plot.

Remember that the linear regression model looks like this: $f(x) = wx + b$. We don't know what the optimal values for w and b are and we want to learn them from data. To do that, we look for such values for w and b that minimize the mean squared error:

$$l \stackrel{\text{def}}{=} \frac{1}{N} \sum_{i=1}^{N} (y_i - (wx_i + b))^2.$$

Gradient descent starts with calculating the partial derivative for every parameter:

$$\frac{\partial l}{\partial w} = \frac{1}{N} \sum_{i=1}^{N} -2x_i(y_i - (wx_i + b));$$

$$\frac{\partial l}{\partial b} = \frac{1}{N} \sum_{i=1}^{N} -2(y_i - (wx_i + b)).$$

(4.1)

To find the partial derivative of the term $(y_i - (wx + b))^2$ with respect to w I applied the *chain rule*. Here, we have the chain $f = f_2(f_1)$ where $f_1 = y_i - (wx + b)$ and $f_2 = f_1^2$. To find a partial derivative of f with respect to w we have to first find the partial derivative of f with respect to f_2 which is equal to $2(y_i - (wx + b))$ (from calculus, we know that the derivative $\frac{\partial}{\partial x} x^2 = 2x$) and then we have to multiply it by the partial derivative of $y_i - (wx + b)$ with respect to w which is equal to $-x$. So overall $\frac{\partial l}{\partial w} = \frac{1}{N} \sum_{i=1}^{N} -2x_i(y_i - (wx_i + b))$. In a similar way, the partial derivative of l with respect to b, $\frac{\partial l}{\partial b}$, was calculated.

Gradient descent proceeds in **epochs**. An epoch consists of using the training set entirely to update each parameter. In the beginning, the first epoch, we initialize[2] $w \leftarrow 0$ and $b \leftarrow 0$. The partial derivatives, $\frac{\partial l}{\partial w}$ and $\frac{\partial l}{\partial b}$ given by eq. 4.1 equal, respectively, $\frac{-2}{N}\sum_{i=1}^{N} x_i y_i$ and $\frac{-2}{N}\sum_{i=1}^{N} y_i$. At each epoch, we update w and b using partial derivatives. The learning rate α controls the size of an update:

$$w \leftarrow w - \alpha \frac{\partial l}{\partial w};$$
$$b \leftarrow b - \alpha \frac{\partial l}{\partial b}.$$

(4.2)

We subtract (as opposed to adding) partial derivatives from the values of parameters because derivatives are indicators of growth of a function. If a derivative is positive at some point[3], then the function grows at this point. Because we want to minimize the objective function, when the derivative is positive we know that we need to move our parameter in the opposite direction (to the left on the axis of coordinates). When the derivative is negative (function is decreasing), we need to move our parameter to the right to decrease the value of the function even more. Subtracting a negative value from a parameter moves it to the right.

At the next epoch, we recalculate partial derivatives using eq. 4.1 with the updated values of w and b; we continue the process until convergence. Typically, we need many epochs until we start seeing that the values for w and b don't change much after each epoch; then we stop.

It's hard to imagine a machine learning engineer who doesn't use Python. So, if you waited for the right moment to start learning Python, this is that moment. Below, I show how to program gradient descent in Python.

The function that updates the parameters w and b during one epoch is shown below:

```
1  def update_w_and_b(spendings, sales, w, b, alpha):
2      dl_dw = 0.0
3      dl_db = 0.0
4      N = len(spendings)
5
6      for i in range(N):
7          dl_dw += -2*spendings[i]*(sales[i] - (w*spendings[i] + b))
8          dl_db += -2*(sales[i] - (w*spendings[i] + b))
9
10     # update w and b
11     w = w - (1/float(N))*dl_dw*alpha
```

---

[2] In complex models, such as neural networks, which have thousands of parameters, the initialization of parameters may significantly affect the solution found using gradient descent. There are different initialization methods (at random, with all zeroes, with small values around zero, and others) and it is an important choice the data analyst has to make.

[3] A point is given by the current values of parameters.

```
12        b = b - (1/float(N))*dl_db*alpha

13

14        return w, b
```

The function that loops over multiple epochs is shown below:

```
15   def train(spendings, sales, w, b, alpha, epochs):
16       for e in range(epochs):
17           w, b = update_w_and_b(spendings, sales, w, b, alpha)
18
19           # log the progress
20           if e % 400 == 0:
21               print("epoch:", e, "loss: ", avg_loss(spendings, sales, w, b))
22
23       return w, b
```

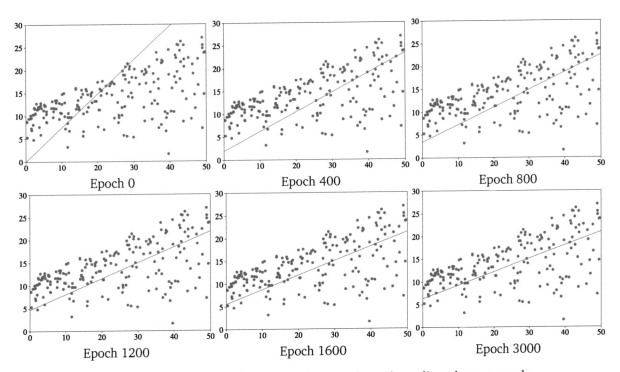

Figure 4.2: The evolution of the regression line through gradient descent epochs.

The function *avg_loss* in the above code snippet is a function that computes the mean squared error. It is defined as:

```
25  def avg_loss(spendings, sales, w, b):
26      N = len(spendings)
27      total_error = 0.0
28      for i in range(N):
29          total_error += (sales[i] - (w*spendings[i] + b))**2
30      return total_error / float(N)
```

If we run the *train* function for $\alpha = 0.001$, $w = 0.0$, $b = 0.0$, and 15,000 epochs, we will see the following output (shown partially):

```
epoch:   0 loss: 92.32078294903626
epoch:   400 loss: 33.79131790081576
epoch:   800 loss: 27.9918542960729
epoch:   1200 loss: 24.33481690722147
epoch:   1600 loss: 22.028754937538633
. . .
epoch:   2800 loss: 19.07940244306619
```

You can see that the average loss decreases as the *train* function loops through epochs. Figure 4.2 shows the evolution of the regression line through epochs.

Finally, once we have found the optimal values of parameters w and b, the only missing piece is a function that makes predictions:

```
31  def predict(x, w, b):
32      return w*x + b
```

Try to execute the following code:

```
33  w, b = train(x, y, 0.0, 0.0, 0.001, 15000)
34  x_new = 23.0
35  y_new = predict(x_new, w, b)
36  print(y_new)
```

The output is 13.97.

Gradient descent is sensitive to the choice of the learning rate α. It is also slow for large datasets. Fortunately, several significant improvements to this algorithm have been proposed.

Minibatch stochastic gradient descent (minibatch SGD) is a version of the algorithm that speeds up the computation by approximating the gradient using smaller batches (subsets) of the training data. SGD itself has various "upgrades". **Adagrad** is a version of SGD that scales α for each parameter according to the history of gradients. As a result, α is reduced for very large gradients and vice-versa. **Momentum** is a method that helps accelerate SGD by orienting the gradient descent in the relevant direction and reducing oscillations. In neural network training, variants of SGD such as **RMSprop** and **Adam**, are very frequently used.

Notice that gradient descent and its variants are not machine learning algorithms. They are solvers of minimization problems in which the function to minimize has a gradient (in most points of its domain).

4.3   How Machine Learning Engineers Work

Unless you are a research scientist or work for a huge corporation with a large R&D budget, you usually don't implement machine learning algorithms yourself. You don't implement gradient descent or some other solver either. You use libraries, most of which are open source. A library is a collection of algorithms and supporting tools implemented with stability and efficiency in mind. The most frequently used in practice open-source machine learning library is *scikit-learn*. It's written in Python and C. Here's how you do linear regression in scikit-learn:

```
1  def train(x, y):
2      from sklearn.linear_model import LinearRegression
3      model = LinearRegression().fit(x,y)
4      return model
5
6  model = train(x,y)
7
8  x_new = 23.0
9  y_new = model.predict(x_new)
10 print(y_new)
```

The output will, again, be 13.97. Easy, right? You can replace LinearRegression with some other type of regression learning algorithm without modifying anything else. It just works. The same can be said about classification. You can easily replace *LogisticRegression* algorithm with *SVC* algorithm (this is scikit-learn's name for the Support Vector Machine algorithm), *DecisionTreeClassifier*, *NearestNeighbors* or many other classification learning algorithms implemented in scikit-learn.

4.4   Learning Algorithms' Particularities

Here, I outline some practical particularities that can differentiate one learning algorithm from another. You already know that different learning algorithms can have different hyperparameters (C in SVM, ϵ and d in ID3). Solvers such as gradient descent can also have hyperparameters, like α for example.

Some algorithms, like decision tree learning, can accept categorical features. For example, if you have a feature "color" that can take values "red", "yellow", or "green", you can keep this feature as is. SVM, logistic and linear regression, as well as kNN (with cosine similarity or Euclidean distance metrics), expect numerical values for all features. All algorithms implemented in scikit-learn expect numerical features. In the next chapter, I show how to convert categorical features into numerical ones.

Some algorithms, like SVM, allow the data analyst to provide weightings for each class. These weightings influence how the decision boundary is drawn. If the weight of some class is high, the learning algorithm tries to not make errors in predicting training examples of this class (typically, for the cost of making an error elsewhere). That could be important if instances of some class are in the minority in your training data, but you would like to avoid misclassifying examples of that class as much as possible.

Some classification models, like SVM and kNN, given a feature vector only output the class. Others, like logistic regression or decision trees, can also return the score between 0 and 1 which can be interpreted as either how confident the model is about the prediction or as the probability that the input example belongs to a certain class[4].

Some classification algorithms (like decision tree learning, logistic regression, or SVM) build the model using the whole dataset at once. If you have got additional labeled examples, you have to rebuild the model from scratch. Other algorithms (such as Naïve Bayes, multilayer perceptron, SGDClassifier/SGDRegressor, PassiveAggressiveClassifier/PassiveAggressiveRegressor in scikit-learn) can be trained iteratively, one batch at a time. Once new training examples are available, you can update the model using only the new data.

Finally, some algorithms, like decision tree learning, SVM, and kNN can be used for both classification and regression, while others can only solve one type of problem: either classification or regression, but not both.

Usually, each library provides the documentation that explains what kind of problem each algorithm solves, what input values are allowed and what kind of output the model produces. The documentation also provides information on hyperparameters.

---

[4]If it's really necessary, the score for SVM and kNN predictions could be synthetically created using simple techniques.

Chapter 5

Basic Practice

Until now, I only mentioned in passing some issues that a data analyst needs to consider when working on a machine learning problem: feature engineering, overfitting, and hyperparameter tuning. In this chapter, we talk about these and other challenges that have to be addressed before you can type model = LogisticRegression().fit(x,y) in scikit-learn.

5.1 Feature Engineering

When a product manager tells you "We need to be able to predict whether a particular customer will stay with us. Here are the logs of customers' interactions with our product for five years." you cannot just grab the data, load it into a library and get a prediction. You need to build a **dataset** first.

Remember from the first chapter that the dataset is the collection of **labeled examples** $\{(\mathbf{x}_i, y_i)\}_{i=1}^{N}$. Each element $\mathbf{x}_i$ among N is called a **feature vector**. A feature vector is a vector in which each dimension $j = 1, \ldots, D$ contains a value that describes the example somehow. That value is called a **feature** and is denoted as $x^{(j)}$.

The problem of transforming raw data into a dataset is called **feature engineering**. For most practical problems, feature engineering is a labor-intensive process that demands from the data analyst a lot of creativity and, preferably, domain knowledge.

For example, to transform the logs of user interaction with a computer system, one could create features that contain information about the user and various statistics extracted from the logs. For each user, one feature would contain the price of the subscription; other features would contain the frequency of connections per day, week and year. Another feature would contain the average session duration in seconds or the average response time for one request, and so on. Everything measurable can be used as a feature. The role of the data analyst is to create *informative* features: those would allow the learning algorithm to build a model that does a good job of predicting labels of the data used for training. Highly informative features are also called features with high *predictive power*. For example, the average duration of a user's session has high predictive power for the problem of predicting whether the user will keep using the application in the future.

We say that a model has a **low bias** when it predicts the training data well. That is, the model makes few mistakes when we use it to predict labels of the examples used to build the model.

5.1.1 One-Hot Encoding

Some learning algorithms only work with numerical feature vectors. When some feature in your dataset is categorical, like "colors" or "days of the week," you can transform such a categorical feature into several binary ones.

If your example has a categorical feature "colors" and this feature has three possible values: "red," "yellow," "green," you can transform this feature into a vector of three numerical values:

$$
\begin{aligned}
\text{red} &= [1, 0, 0] \\
\text{yellow} &= [0, 1, 0] \\
\text{green} &= [0, 0, 1]
\end{aligned}
\tag{5.1}
$$

By doing so, you increase the dimensionality of your feature vectors. You should not transform red into 1, yellow into 2, and green into 3 to avoid increasing the dimensionality because that would imply that there's an order among the values in this category and this specific order is important for the decision making. If the order of a feature's values is not important, using ordered numbers as values is likely to confuse the learning algorithm,[1] because the algorithm will try to find a regularity where there's no one, which may potentially lead to overfitting.

5.1.2 Binning

An opposite situation, occurring less frequently in practice, is when you have a numerical feature but you want to convert it into a categorical one. **Binning** (also called **bucketing**) is the process of converting a continuous feature into multiple binary features called bins or buckets, typically based on value range. For example, instead of representing age as a single real-valued feature, the analyst could chop ranges of age into discrete bins: all ages between 0 and 5 years-old could be put into one bin, 6 to 10 years-old could be in the second bin, 11 to 15 years-old could be in the third bin, and so on.

---

[1]When the ordering of values of some categorical variable matters, we can replace those values by numbers by keeping only one variable. For example, if our variable represents the quality of an article, and the values are $\{poor, decent, good, excellent\}$, then we could replace those categories by numbers, for example, $\{1, 2, 3, 4\}$.

Let feature $j = 4$ represent age. By applying binning, we replace this feature with the corresponding bins. Let the three new bins, "age_bin1", "age_bin2" and "age_bin3" be added with indexes $j = 123$, $j = 124$ and $j = 125$ respectively (by default the values of these three new features are 0). Now if $x_i^{(4)} = 7$ for some example $\mathbf{x}_i$, then we set feature $x_i^{(124)}$ to 1; if $x_i^{(4)} = 13$, then we set feature $x_i^{(125)}$ to 1, and so on.

In some cases, a carefully designed binning can help the learning algorithm to learn using fewer examples. It happens because we give a "hint" to the learning algorithm that if the value of a feature falls within a specific range, the exact value of the feature doesn't matter.

5.1.3 Normalization

Normalization is the process of converting an actual range of values which a numerical feature can take, into a standard range of values, typically in the interval $[-1, 1]$ or $[0, 1]$.

For example, suppose the natural range of a particular feature is 350 to 1450. By subtracting 350 from every value of the feature, and dividing the result by 1100, one can normalize those values into the range $[0, 1]$.

More generally, the normalization formula looks like this:

$$\bar{x}^{(j)} = \frac{x^{(j)} - min^{(j)}}{max^{(j)} - min^{(j)}},$$

where $min^{(j)}$ and $max^{(j)}$ are, respectively, the minimum and the maximum value of the feature j in the dataset.

Why do we normalize? Normalizing the data is not a strict requirement. However, in practice, it can lead to an increased speed of learning. Remember the gradient descent example from the previous chapter. Imagine you have a two-dimensional feature vector. When you update the parameters of $w^{(1)}$ and $w^{(2)}$, you use partial derivatives of the mean squared error with respect to $w^{(1)}$ and $w^{(2)}$. If $x^{(1)}$ is in the range $[0, 1000]$ and $x^{(2)}$ the range $[0, 0.0001]$, then the derivative with respect to a larger feature will dominate the update.

Additionally, it's useful to ensure that our inputs are roughly in the same relatively small range to avoid problems which computers have when working with very small or very big numbers (known as numerical overflow).

5.1.4 Standardization

Standardization (or **z-score normalization**) is the procedure during which the feature values are rescaled so that they have the properties of a *standard normal distribution* with $\mu = 0$ and $\sigma = 1$, where μ is the mean (the average value of the feature, averaged over all examples in the dataset) and σ is the standard deviation from the mean.

Standard scores (or z-scores) of features are calculated as follows:

$$\hat{x}^{(j)} = \frac{x^{(j)} - \mu^{(j)}}{\sigma^{(j)}}.$$

You may ask when you should use normalization and when standardization. There's no definitive answer to this question. Usually, if your dataset is not too big and you have time, you can try both and see which one performs better for your task.

If you don't have time to run multiple experiments, as a rule of thumb:

- unsupervised learning algorithms, in practice, more often benefit from standardization than from normalization;
- standardization is also preferred for a feature if the values this feature takes are distributed close to a normal distribution (so-called bell curve);
- again, standardization is preferred for a feature if it can sometimes have extremely high or low values (outliers); this is because normalization will "squeeze" the normal values into a very small range;
- in all other cases, normalization is preferable.

Feature rescaling is usually beneficial to most learning algorithms. However, modern implementations of the learning algorithms, which you can find in popular libraries, are robust to features lying in different ranges.

5.1.5 Dealing with Missing Features

In some cases, the data comes to the analyst in the form of a dataset with features already defined. In some examples, values of some features can be missing. That often happens when the dataset was handcrafted, and the person working on it forgot to fill some values or didn't get them measured at all.

The typical approaches of dealing with missing values for a feature include:

- removing the examples with missing features from the dataset (that can be done if your dataset is big enough so you can sacrifice some training examples);
- using a learning algorithm that can deal with missing feature values (depends on the library and a specific implementation of the algorithm);
- using a **data imputation** technique.

5.1.6 Data Imputation Techniques

One data imputation technique consists in replacing the missing value of a feature by an average value of this feature in the dataset:

$$\hat{x}^{(j)} \leftarrow \frac{1}{N} x^{(j)}.$$

Another technique is to replace the missing value with a value outside the normal range of values. For example, if the normal range is $[0, 1]$, then you can set the missing value to 2 or -1. The idea is that the learning algorithm will learn what is best to do when the feature has a value significantly different from regular values. Alternatively, you can replace the missing value by a value in the middle of the range. For example, if the range for a feature is $[-1, 1]$, you can set the missing value to be equal to 0. Here, the idea is that the value in the middle of the range will not significantly affect the prediction.

A more advanced technique is to use the missing value as the target variable for a regression problem. You can use all remaining features $[x_i^{(1)}, x_i^{(2)}, \ldots, x_i^{(j-1)}, x_i^{(j+1)}, \ldots, x_i^{(D)}]$ to form a feature vector $\hat{x}_i$, set $\hat{y}_i \leftarrow x^{(j)}$, where j is the feature with a missing value. Then you build a regression model to predict $\hat{y}$ from $\hat{x}$. Of course, to build training examples $(\hat{x}, \hat{y})$, you only use those examples from the original dataset, in which the value of feature j is present.

Finally, if you have a significantly large dataset and just a few features with missing values, you can increase the dimensionality of your feature vectors by adding a binary indicator feature for each feature with missing values. Let's say feature $j = 12$ in your D-dimensional dataset has missing values. For each feature vector x, you then add the feature $j = D + 1$ which is equal to 1 if the value of feature 12 is present in x and 0 otherwise. The missing feature value then can be replaced by 0 or any number of your choice.

At prediction time, if your example is not complete, you should use the same data imputation technique to fill the missing features as the technique you used to complete the training data.

Before you start working on the learning problem, you cannot tell which data imputation technique will work the best. Try several techniques, build several models and select the one that works the best.

5.2   Learning Algorithm Selection

Choosing a machine learning algorithm can be a difficult task. If you have much time, you can try all of them. However, usually the time you have to solve a problem is limited. You can ask yourself several questions before starting to work on the problem. Depending on your answers, you can shortlist some algorithms and try them on your data.

- Explainability

Does your model have to be explainable to a non-technical audience? Most very accurate learning algorithms are so-called "black boxes." They learn models that make very few errors, but why a model made a specific prediction could be very hard to understand and even harder to explain. Examples of such models are neural networks or ensemble models.

On the other hand, kNN, linear regression, or decision tree learning algorithms produce models that are not always the most accurate, however, the way they make their prediction is very straightforward.

- In-memory vs. out-of-memory

Can your dataset be fully loaded into the RAM of your server or personal computer? If yes, then you can choose from a wide variety of algorithms. Otherwise, you would prefer **incremental learning algorithms** that can improve the model by adding more data gradually.

- Number of features and examples

How many training examples do you have in your dataset? How many features does each example have? Some algorithms, including **neural networks** and **gradient boosting** (we consider both later), can handle a huge number of examples and millions of features. Others, like SVM, can be very modest in their capacity.

- Categorical vs. numerical features

Is your data composed of categorical only, or numerical only features, or a mix of both? Depending on your answer, some algorithms cannot handle your dataset directly, and you would need to convert your categorical features into numerical ones.

- Nonlinearity of the data

Is your data linearly separable or can it be modeled using a linear model? If yes, SVM with the linear kernel, logistic or linear regression can be good choices. Otherwise, deep neural networks or ensemble algorithms, discussed in Chapters 6 and 7, might work better.

- Training speed

How much time is a learning algorithm allowed to use to build a model? Neural networks are known to be slow to train. Simple algorithms like logistic and linear regression or decision trees are much faster. Specialized libraries contain very efficient implementations of some algorithms; you may prefer to do research online to find such libraries. Some algorithms, such as random forests, benefit from the availability of multiple CPU cores, so their model building time can be significantly reduced on a machine with dozens of cores.

- Prediction speed

How fast does the model have to be when generating predictions? Will your model be used in production where very high throughput is required? Algorithms like SVMs, linear and logistic regression, and (some types of) neural networks, are extremely fast at the prediction time. Others, like kNN, ensemble algorithms, and very deep or recurrent neural networks, are slower[2].

If you don't want to guess the best algorithm for your data, a popular way to choose one is by testing it on the **validation set**. We talk about that below. Alternatively, if you use scikit-learn, you could try their algorithm selection diagram shown in Figure 5.1.

5.3   Three Sets

Until now, I used the expressions "dataset" and "training set" interchangeably. However, in practice data analysts work with three distinct sets of labeled examples:

1) training set,
2) validation set, and
3) test set.

Once you have got your annotated dataset, the first thing you do is you shuffle the examples and split the dataset into three subsets: **training**, **validation**, and **test**. The training set is usually the biggest one; you use it to build the model. The validation and test sets are roughly the same sizes, much smaller than the size of the training set. The learning algorithm *cannot* use examples from these two subsets to build the model. That is why those two sets are often called **holdout sets**.

There's no optimal proportion to split the dataset into these three subsets. In the past, the rule of thumb was to use 70% of the dataset for training, 15% for validation and 15% for testing. However, in the age of big data, datasets often have millions of examples. In such cases, it could be reasonable to keep 95% for training and 2.5%/2.5% for validation/testing.

You may wonder, what is the reason to have three sets and not one. The answer is simple: when we build a model, what we do not want is for the model to only do well at predicting labels of examples the learning algorithms has already seen. A trivial algorithm that simply memorizes all training examples and then uses the memory to "predict" their labels will make no mistakes when asked to predict the labels of the training examples, but such an algorithm would be useless in practice. What we really want is a model that is good at predicting examples that the learning algorithm didn't see: we want good performance on a holdout set.

---

[2]The prediction speed of kNN and ensemble methods implemented in the modern libraries are still pretty fast. Don't be afraid of using these algorithms in your practice.

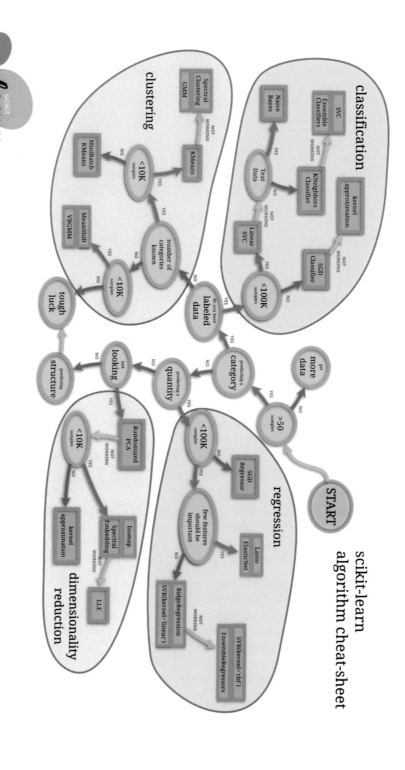

Figure 5.1: Machine learning algorithm selection diagram for scikit-learn.

Why do we need two holdout sets and not one? We use the validation set to 1) choose the learning algorithm and 2) find the best values of hyperparameters. We use the test set to assess the model before delivering it to the client or putting it in production.

5.4 Underfitting and Overfitting

I mentioned above the notion of **bias**. I said that a model has a low bias if it predicts well the labels of the training data. If the model makes many mistakes on the training data, we say that the model has a **high bias** or that the model **underfits**. So, underfitting is the inability of the model to predict well the labels of the data it was trained on. There could be several reasons for underfitting, the most important of which are:

- your model is too simple for the data (for example a linear model can often underfit);
- the features you engineered are not informative enough.

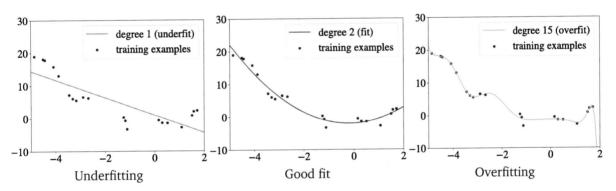

Figure 5.2: Examples of underfitting (linear model), good fit (quadratic model), and overfitting (polynomial of degree 15).

The first reason is easy to illustrate in the case of one-dimensional regression: the dataset can resemble a curved line, but our model is a straight line. The second reason can be illustrated like this: let's say you want to predict whether a patient has cancer, and the features you have are height, blood pressure, and heart rate. These three features are clearly not good predictors for cancer so our model will not be able to learn a meaningful relationship between these features and the label.

The solution to the problem of underfitting is to try a more complex model or to engineer features with higher predictive power.

Overfitting is another problem a model can exhibit. The model that overfits predicts very well the training data but poorly the data from at least one of the two holdout sets. I already gave an illustration of overfitting in Chapter 3. Several reasons can lead to overfitting, the most important of which are:

- your model is too complex for the data (for example a very tall decision tree or a very deep or wide neural network often overfit);
- you have too many features but a small number of training examples.

In the literature, you can find another name for the problem of overfitting: the problem of **high variance**. This term comes from statistics. The variance is an error of the model due to its sensitivity to small fluctuations in the training set. It means that if your training data was sampled differently, the learning would result in a significantly different model. Which is why the model that overfits performs poorly on the test data: test and training data are sampled from the dataset independently of one another.

Even the simplest model, such as linear, can overfit the data. That usually happens when the data is high-dimensional, but the number of training examples is relatively low. In fact, when feature vectors are very high-dimensional, the linear learning algorithm can build a model that assigns non-zero values to most parameters $w^{(j)}$ in the parameter vector $\mathbf{w}$, trying to find very complex relationships between all available features to predict labels of training examples perfectly.

Such a complex model will most likely predict poorly the labels of the holdout examples. This is because by trying to perfectly predict labels of all training examples, the model will also learn the idiosyncrasies of the training set: the noise in the values of features of the training examples, the sampling imperfection due to the small dataset size, and other artifacts extrinsic to the decision problem at hand but present in the training set.

Figure 5.2 illustrates a one-dimensional dataset for which a regression model underfits, fits well and overfits the data.

Several solutions to the problem of overfitting are possible:

1. Try a simpler model (linear instead of polynomial regression, or SVM with a linear kernel instead of RBF, a neural network with fewer layers/units).
2. Reduce the dimensionality of examples in the dataset (for example, by using one of the dimensionality reduction techniques discussed in Chapter 9).
3. Add more training data, if possible.
4. Regularize the model.

Regularization is the most widely used approach to prevent overfitting.

5.5 Regularization

Regularization is an umbrella term that encompasses methods that force the learning algorithm to build a less complex model. In practice, that often leads to slightly higher bias but significantly reduces the variance. This problem is known in the literature as the **bias-variance tradeoff**.

The two most widely used types of regularization are called **L1** and **L2 regularization**. The idea is quite simple. To create a regularized model, we modify the objective function by adding a penalizing term whose value is higher when the model is more complex.

For simplicity, I illustrate regularization using the example of linear regression. The same principle can be applied to a wide variety of models.

Recall the linear regression objective:

$$\min_{\mathbf{w},b} \frac{1}{N} \sum_{i=1}^{N} (f_{\mathbf{w},b}(\mathbf{x}_i) - y_i)^2. \tag{5.2}$$

An L1-regularized objective looks like this:

$$\min_{\mathbf{w},b} \left[C|\mathbf{w}| + \frac{1}{N} \sum_{i=1}^{N} (f_{\mathbf{w},b}(\mathbf{x}_i) - y_i)^2 \right], \tag{5.3}$$

where $|\mathbf{w}| \stackrel{\text{def}}{=} \sum_{j=1}^{D} |w^{(j)}|$ and C is a hyperparameter that controls the importance of regularization. If we set C to zero, the model becomes a standard non-regularized linear regression model. On the other hand, if we set to C to a high value, the learning algorithm will try to set most $w^{(j)}$ to a very small value or zero to minimize the objective, and the model will become very simple which can lead to underfitting. Your role as the data analyst is to find such a value of the hyperparameter C that doesn't increase the bias too much but reduces the variance to a level reasonable for the problem at hand. In the next section, I will show how to do that.

An L2-regularized objective looks like this:

$$\min_{\mathbf{w},b} \left[C\|\mathbf{w}\|^2 + \frac{1}{N} \sum_{i=1}^{N} (f_{\mathbf{w},b}(\mathbf{x}_i) - y_i)^2 \right], \quad \text{where } \|\mathbf{w}\|^2 \stackrel{\text{def}}{=} \sum_{j=1}^{D} (w^{(j)})^2. \tag{5.4}$$

In practice, L1 regularization produces a **sparse model**, a model that has most of its parameters (in case of linear models, most of $w^{(j)}$) equal to zero, provided the hyperparameter C is large enough. So L1 performs **feature selection** by deciding which features are essential for prediction and which are not. That can be useful in case you want to increase model explainability. However, if your only goal is to maximize the performance of the model on the holdout data, then L2 usually gives better results. L2 also has the advantage of being differentiable, so gradient descent can be used for optimizing the objective function.

L1 and L2 regularization methods were also combined in what is called **elastic net regularization** with L1 and L2 regularizations being special cases. You can find in the literature the name **ridge regularization** for L2 and **lasso** for L1.

In addition to being widely used with linear models, L1 and L2 regularization are also frequently used with neural networks and many other types of models, which directly minimize an objective function.

Neural networks also benefit from two other regularization techniques: **dropout** and **batch-normalization**. There are also non-mathematical methods that have a regularization effect: **data augmentation** and **early stopping**. We talk about these techniques in Chapter 8.

5.6   Model Performance Assessment

Once you have a model which our learning algorithm has built using the training set, how can you say how good the model is? You use the test set to assess the model.

The test set contains the examples that the learning algorithm has never seen before, so if our model performs well on predicting the labels of the examples from the test set, we say that our model **generalizes well** or, simply, that it's good.

To be more rigorous, machine learning specialists use various formal metrics and tools to assess the model performance. For regression, the assessment of the model is quite simple. A well-fitting regression model results in predicted values close to the observed data values. The **mean model**, which always predicts the average of the labels in the training data, generally would be used if there were no informative features. The fit of a regression model being assessed should, therefore, be better than the fit of the mean model. If this is the case, then the next step is to compare the performances of the model on the training and the test data.

To do that, we compute the mean squared error[3] (MSE) for the training, and, separately, for the test data. If the MSE of the model on the test data is *substantially higher* than the MSE obtained on the training data, this is a sign of overfitting. Regularization or a better hyperparameter tuning could solve the problem. The meaning of "substantially higher" depends on the problem at hand and has to be decided by the data analyst jointly with the decision maker/product owner who ordered the model.

For classification, things are a little bit more complicated. The most widely used metrics and tools to assess the classification model are:

- confusion matrix,
- accuracy,
- cost-sensitive accuracy,
- precision/recall, and
- area under the ROC curve.

To simplify the illustration, I use a binary classification problem. Where necessary, I show how to extend the approach to the multiclass case.

---

[3]Or any other type of average loss function that makes sense.

5.6.1 Confusion Matrix

The **confusion matrix** is a table that summarizes how successful the classification model is at predicting examples belonging to various classes. One axis of the confusion matrix is the label that the model predicted, and the other axis is the actual label. In a binary classification problem, there are two classes. Let's say, the model predicts two classes: "spam" and "not_spam":

	spam (predicted)	not_spam (predicted)
spam (actual)	23 (TP)	1 (FN)
not_spam (actual)	12 (FP)	556 (TN)

The above confusion matrix shows that of the 24 examples that actually were spam, the model correctly classified 23 as spam. In this case, we say that we have 23 **true positives** or TP = 23. The model incorrectly classified 1 example as not_spam. In this case, we have 1 **false negative**, or FN = 1. Similarly, of 568 examples that actually were not spam, 556 were correctly classified (556 **true negatives** or TN = 556), and 12 were incorrectly classified (12 **false positives**, FP = 12).

The confusion matrix for multiclass classification has as many rows and columns as there are different classes. It can help you to determine mistake patterns. For example, a confusion matrix could reveal that a model trained to recognize different species of animals tends to mistakenly predict "cat" instead of "panther," or "mouse" instead of "rat." In this case, you can decide to add more labeled examples of these species to help the learning algorithm to "see" the difference between them. Alternatively, you might add additional features the learning algorithm can use to build a model that would better distinguish between these species.

Confusion matrix is used to calculate two other performance metrics: **precision** and **recall**.

5.6.2 Precision/Recall

The two most frequently used metrics to assess the model are **precision** and **recall**. Precision is the ratio of correct positive predictions to the overall number of positive predictions:

$$\text{precision} \overset{\text{def}}{=} \frac{\text{TP}}{\text{TP} + \text{FP}}.$$

Recall is the ratio of correct positive predictions to the overall number of positive examples in the dataset:

$$\text{recall} \overset{\text{def}}{=} \frac{\text{TP}}{\text{TP} + \text{FN}}.$$

To understand the meaning and importance of precision and recall for the model assessment it is often useful to think about the prediction problem as the problem of research of documents in the database using a query. The precision is the proportion of relevant documents in the list of all returned documents. The recall is the ratio of the relevant documents returned by the search engine to the total number of the relevant documents that could have been returned.

In the case of the spam detection problem, we want to have high precision (we want to avoid making mistakes by detecting that a legitimate message is spam) and we are ready to tolerate lower recall (we tolerate some spam messages in our inbox).

Almost always, in practice, we have to choose between a high precision or a high recall. It's usually impossible to have both. We can achieve either of the two by various means:

- by assigning a higher weighting to the examples of a specific class (the SVM algorithm accepts weightings of classes as input);
- by tuning hyperparameters to maximize precision or recall on the validation set;
- by varying the decision threshold for algorithms that return probabilities of classes; for instance, if we use logistic regression or decision tree, to increase precision (at the cost of a lower recall), we can decide that the prediction will be positive only if the probability returned by the model is higher than 0.9.

Even if precision and recall are defined for the binary classification case, you can always use it to assess a multiclass classification model. To do that, first select a class for which you want to assess these metrics. Then you consider all examples of the selected class as positives and all examples of the remaining classes as negatives.

5.6.3 Accuracy

Accuracy is given by the number of correctly classified examples divided by the total number of classified examples. In terms of the confusion matrix, it is given by:

$$\text{accuracy} \overset{\text{def}}{=} \frac{\text{TP} + \text{TN}}{\text{TP} + \text{TN} + \text{FP} + \text{FN}}. \tag{5.5}$$

Accuracy is a useful metric when errors in predicting all classes are equally important. In case of the spam/not spam, this may not be the case. For example, you would tolerate false positives less than false negatives. A false positive in spam detection is the situation in which your friend sends you an email, but the model labels it as spam and doesn't show you. On the other hand, the false negative is less of a problem: if your model doesn't detect a small percentage of spam messages, it's not a big deal.

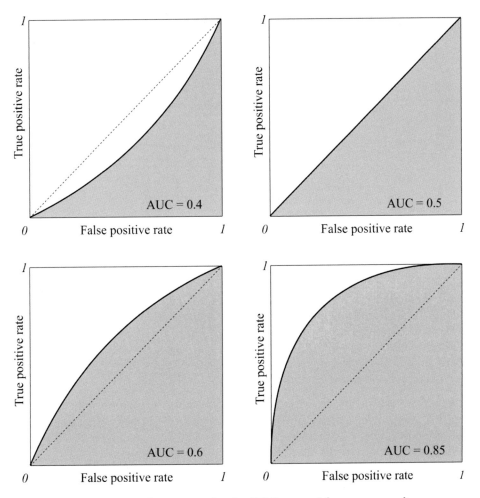

Figure 5.3: The area under the ROC curve (shown on gray).

5.6.4 Cost-Sensitive Accuracy

For dealing with the situation in which different classes have different importance, a useful metric is **cost-sensitive accuracy**. To compute a cost-sensitive accuracy, you first assign a cost (a positive number) to both types of mistakes: FP and FN. You then compute the counts TP, TN, FP, FN as usual and multiply the counts for FP and FN by the corresponding cost before calculating the accuracy using eq. 5.5.

5.6.5   Area under the ROC Curve (AUC)

The ROC curve (stands for "receiver operating characteristic;" the term comes from radar engineering) is a commonly used method to assess the performance of classification models. ROC curves use a combination of the **true positive rate** (defined exactly as **recall**) and false positive rate (the proportion of negative examples predicted incorrectly) to build up a summary picture of the classification performance.

The true positive rate (TPR) and the false positive rate (FPR) are respectively defined as,

$$\text{TPR} \overset{\text{def}}{=} \frac{\text{TP}}{\text{TP} + \text{FN}} \quad \text{and} \quad \text{FPR} \overset{\text{def}}{=} \frac{\text{FP}}{\text{FP} + \text{TN}}.$$

ROC curves can only be used to assess classifiers that return some confidence score (or a probability) of prediction. For example, logistic regression, neural networks, and decision trees (and ensemble models based on decision trees) can be assessed using ROC curves.

To draw a ROC curve, you first discretize the range of the confidence score. If this range for a model is $[0, 1]$, then you can discretize it like this: $[0, 0.1, 0.2, 0.3, 0.4, 0.5, 0.6, 0.7, 0.8, 0.9, 1]$. Then, you use each discrete value as the prediction threshold and predict the labels of examples in your dataset using the model and this threshold. For example, if you want to compute TPR and FPR for the threshold equal to 0.7, you apply the model to each example, get the score, and, if the score is higher than or equal to 0.7, you predict the positive class; otherwise, you predict the negative class.

Look at the illustration in Figure 5.3. It's easy to see that if the threshold is 0, all our predictions will be positive, so both TPR and FPR will be 1 (the upper right corner). On the other hand, if the threshold is 1, then no positive prediction will be made, both TPR and FPR will be 0 which corresponds to the lower left corner.

The higher the **area under the ROC curve** (AUC), the better the classifier. A classifier with an AUC higher than 0.5 is better than a random classifier. If AUC is lower than 0.5, then something is wrong with your model. A perfect classifier would have an AUC of 1. Usually, if your model behaves well, you obtain a good classifier by selecting the value of the threshold that gives TPR close to 1 while keeping FPR near 0.

ROC curves are popular because they are relatively simple to understand, they capture more than one aspect of the classification (by taking both false positives and negatives into account) and allow visually and with low effort comparing the performance of different models.

5.7   Hyperparameter Tuning

When I presented learning algorithms, I mentioned that you as a data analyst have to select good values for the algorithm's hyperparameters, such as ϵ and d for ID3, C for SVM, or α for gradient descent. But what does that exactly mean? Which value is the best and how to find it? In this section, I answer these essential questions.

As you already know, hyperparameters aren't optimized by the learning algorithm itself. The data analyst has to "tune" hyperparameters by experimentally finding the best combination of values, one per hyperparameter.

One typical way to do that, when you have enough data to have a decent validation set (in which each class is represented by at least a couple of dozen examples) and the number of hyperparameters and their range is not too large is to use **grid search**.

Grid search is the most simple **hyperparameter tuning** technique. Let's say you train an SVM and you have two hyperparameters to tune: the penalty parameter C (a positive real number) and the kernel (either "linear" or "rbf").

If it's the first time you are working with this particular dataset, you don't know what is the possible range of values for C. The most common trick is to use a logarithmic scale. For example, for C you can try the following values: [0.001, 0.01, 0.1, 1, 10, 100, 1000]. In this case you have 14 combinations of hyperparameters to try: [(0.001, "linear"), (0.01, "linear"), (0.1, "linear"), (1, "linear"), (10, "linear"), (100, "linear"), (1000, "linear"), (0.001, "rbf"), (0.01, "rbf"), (0.1, "rbf"), (1, "rbf"), (10, "rbf"), (100, "rbf"), (1000, "rbf")].

You use the training set and train 14 models, one for each combination of hyperparameters. Then you assess the performance of each model on the validation data using one of the metrics we discussed in the previous section (or some other metric that matters to you). Finally, you keep the model that performs the best according to the metric.

Once the best pair of hyperparameters is found, you can try to explore the values close to the best ones in some region around them. Sometimes, this can result in an even better model.

Finally, you assess the selected model using the test set.

As you could notice, trying all combinations of hyperparameters, especially if there are more than a couple of them, could be time-consuming, especially for large datasets. There are more efficient techniques, such as **random search** and **Bayesian hyperparameter optimization**.

Random search differs from grid search in that you no longer provide a discrete set of values to explore for each hyperparameter; instead, you provide a statistical distribution for each hyperparameter from which values are randomly sampled and set the total number of combinations you want to try.

Bayesian techniques differ from random or grid search in that they use past evaluation results to choose the next values to evaluate. The idea is to limit the number of expensive optimizations of the objective function by choosing the next hyperparameter values based on those that have done well in the past.

There are also **gradient-based techniques, evolutionary optimization techniques,** and other algorithmic hyperparameter tuning techniques. Most modern machine learning libraries implement one or more such techniques. There are also hyperparameter tuning libraries that can help you to tune hyperparameters of virtually any learning algorithm, including ones you programmed yourself.

5.7.1 Cross-Validation

When you don't have a decent validation set to tune your hyperparameters on, the common technique that can help you is called **cross-validation**. When you have few training examples, it could be prohibitive to have both validation and test set. You would prefer to use more data to train the model. In such a case, you only split your data into a training and a test set. Then you use cross-validation on the training set to simulate a validation set.

Cross-validation works as follows. First, you fix the values of the hyperparameters you want to evaluate. Then you split your training set into several subsets of the same size. Each subset is called a *fold*. Typically, five-fold cross-validation is used in practice. With five-fold cross-validation, you randomly split your training data into five folds: $\{F_1, F_2, \ldots, F_5\}$. Each F_k, $k = 1, \ldots, 5$ contains 20% of your training data. Then you train five models as follows. To train the first model, f_1, you use all examples from folds F_2, F_3, F_4, and F_5 as the training set and the examples from F_1 as the validation set. To train the second model, f_2, you use the examples from folds F_1, F_3, F_4, and F_5 to train and the examples from F_2 as the validation set. You continue building models iteratively like this and compute the value of the metric of interest on each validation set, from F_1 to F_5. Then you average the five values of the metric to get the final value.

You can use grid search with cross-validation to find the best values of hyperparameters for your model. Once you have found these values, you use the entire training set to build the model with these best values of hyperparameters you have found via cross-validation. Finally, you assess the model using the test set.

Chapter 6

Neural Networks and Deep Learning

First of all, you already know what a neural network is, and you already know how to build such a model. Yes, it's logistic regression! As a matter of fact, the logistic regression model, or rather its generalization for multiclass classification, called the softmax regression model, is a standard unit in a neural network.

6.1 Neural Networks

If you understood linear regression, logistic regression, and gradient descent, understanding neural networks should not be a problem.

A **neural network** (NN), just like a regression or an SVM model, is a mathematical function:

$$y = f_{NN}(\mathbf{x}).$$

The function f_{NN} has a particular form: it's a **nested function**. You have probably already heard of neural network **layers**. So, for a 3-layer neural network that returns a scalar, f_{NN} looks like this:

$$y = f_{NN}(\mathbf{x}) = f_3(\boldsymbol{f}_2(\boldsymbol{f}_1(\mathbf{x}))).$$

In the above equation, $\boldsymbol{f}_1$ and $\boldsymbol{f}_2$ are vector functions of the following form:

$$\boldsymbol{f}_l(\mathbf{z}) \stackrel{\text{def}}{=} \boldsymbol{g}_l(\mathbf{W}_l\mathbf{z} + \mathbf{b}_l), \tag{6.1}$$

61

where l is called the layer index and can span from 1 to any number of layers. The function g_l is called an **activation function**. It is a fixed, usually nonlinear function chosen by the data analyst before the learning is started. The parameters $\mathbf{W}_l$ (a matrix) and $\mathbf{b}_l$ (a vector) for each layer are learned using the familiar gradient descent by optimizing, depending on the task, a particular cost function (such as MSE). Compare eq. 6.1 with the equation for logistic regression, where you replace g_l by the sigmoid function, and you will not see any difference. The function f_3 is a scalar function for the regression task, but can also be a vector function depending on your problem.

You may probably wonder why a matrix $\mathbf{W}_l$ is used and not a vector $\mathbf{w}_l$. The reason is that g_l is a vector function. Each row $\mathbf{w}_{l,u}$ (u for unit) of the matrix $\mathbf{W}_l$ is a vector of the same dimensionality as $\mathbf{z}$. Let $a_{l,u} = \mathbf{w}_{l,u}\mathbf{z} + b_{l,u}$. The output of $f_l(\mathbf{z})$ is a vector $[g_l(a_{l,1}), g_l(a_{l,2}), \dots, g_l(a_{l,size_l})]$, where g_l is some scalar function[1], and $size_l$ is the number of units in layer l. To make it more concrete, let's consider one architecture of neural networks called **multilayer perceptron** and often referred to as a **vanilla neural network**.

6.1.1 Multilayer Perceptron Example

We have a closer look at one particular configuration of neural networks called **feed-forward neural networks** (FFNN), and more specifically the architecture called a **multilayer percep-tron** (MLP). As an illustration, let's consider an MLP with three layers. Our network takes a two-dimensional feature vector as input and outputs a number. This FFNN can be a regression or a classification model, depending on the activation function used in the third, output layer.

Our MLP is depicted in Figure 6.1. The neural network is represented graphically as a connected combination of **units** logically organized into one or more **layers**. Each unit is represented by either a circle or a rectangle. The inbound arrow represents an input of a unit and indicates where this input came from. The outbound arrow indicates the output of a unit.

The output of each unit is the result of the mathematical operation written inside the rectangle. Circle units don't do anything with the input; they just send their input directly to the output.

The following happens in each rectangle unit. Firstly, all inputs of the unit are joined together to form an input vector. Then the unit applies a linear transformation to the input vector, exactly like linear regression model does with its input feature vector. Finally, the unit applies an activation function g to the result of the linear transformation and obtains the output value, a real number. In a vanilla FFNN, the output value of a unit of some layer becomes an input value of each of the units of the subsequent layer.

In Figure 6.1, the activation function g_l has one index: l, the index of the layer the unit belongs to. Usually, all units of a layer use the same activation function, but it's not a rule. Each layer can have a different number of units. Each unit has its parameters $\mathbf{w}_{l,u}$ and $b_{l,u}$, where u is the index of the unit, and l is the index of the layer. The vector $\mathbf{y}_{l-1}$ in each unit is defined as $[y_{l-1}^{(1)}, y_{l-1}^{(2)}, y_{l-1}^{(3)}, y_{l-1}^{(4)}]$. The vector $\mathbf{x}$ in the first layer is defined as $[x^{(1)}, \dots, x^{(D)}]$.

---

[1] A scalar function outputs a scalar, that is a simple number and not a vector.

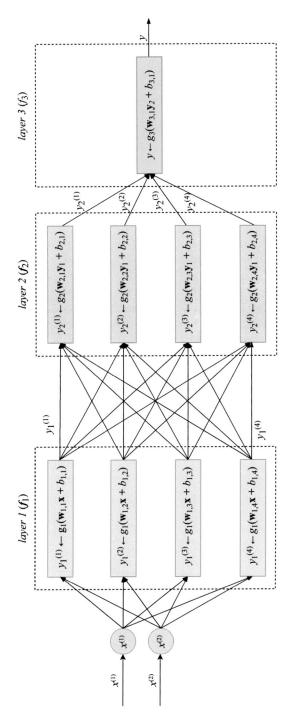

Figure 6.1: A multilayer perceptron with two-dimensional input, two layers with four units and one output layer with one unit.

As you can see in Figure 6.1, in multilayer perceptron all outputs of one layer are connected to each input of the succeeding layer. This architecture is called **fully-connected**. A neural network can contain **fully-connected layers**. Those are the layers whose units receive as inputs the outputs of each of the units of the previous layer.

6.1.2   Feed-Forward Neural Network Architecture

If we want to solve a regression or a classification problem discussed in previous chapters, the last (the rightmost) layer of a neural network usually contains only one unit. If the activation function g_{last} of the last unit is linear, then the neural network is a regression model. If the g_{last} is a logistic function, the neural network is a binary classification model.

The data analyst can choose any mathematical function as $g_{l,u}$, assuming it's differentiable[2]. The latter property is essential for gradient descent used to find the values of the parameters $\mathbf{w}_{l,u}$ and $b_{l,u}$ for all l and u. The primary purpose of having nonlinear components in the function f_{NN} is to allow the neural network to approximate nonlinear functions. Without nonlinearities, f_{NN} would be linear, no matter how many layers it has. The reason is that $\mathbf{W}_l\mathbf{z} + \mathbf{b}_l$ is a linear function and a linear function of a linear function is also linear.

Popular choices of activation functions are the logistic function, already known to you, as well as **TanH** and **ReLU**. The former is the hyperbolic tangent function, similar to the logistic function but ranging from -1 to 1 (without reaching them). The latter is the rectified linear unit function, which equals to zero when its input z is negative and to z otherwise:

$$tanh(z) = \frac{e^z - e^{-z}}{e^z + e^{-z}},$$

$$relu(z) = \begin{cases} 0 & \text{if } z < 0 \\ z & \text{otherwise} \end{cases}.$$

As I said above, $\mathbf{W}_l$ in the expression $\mathbf{W}_l\mathbf{z} + \mathbf{b}_l$, is a matrix, while $\mathbf{b}_l$ is a vector. That looks different from linear regression's $\mathbf{w}\mathbf{z} + b$. In matrix $\mathbf{W}_l$, each row u corresponds to a vector of parameters $\mathbf{w}_{l,u}$. The dimensionality of the vector $\mathbf{w}_{l,u}$ equals to the number of units in the layer $l - 1$. The operation $\mathbf{W}_l\mathbf{z}$ results in a vector $\mathbf{a}_l \stackrel{\text{def}}{=} [\mathbf{w}_{l,1}\mathbf{z}, \mathbf{w}_{l,2}\mathbf{z}, \dots, \mathbf{w}_{l,size_l}\mathbf{z}]$. Then the sum $\mathbf{a}_l + \mathbf{b}_l$ gives a $size_l$-dimensional vector $\mathbf{c}_l$. Finally, the function $g_l(\mathbf{c}_l)$ produces the vector $\mathbf{y}_l \stackrel{\text{def}}{=} [y_l^{(1)}, y_l^{(2)}, \dots, y_l^{(size_l)}]$ as output.

---

[2]The function has to be differentiable across its whole domain or in the majority of the points of its domain. For example, ReLU is not differentiable at 0.

6.2 Deep Learning

Deep learning refers to training neural networks with more than two non-output layers. In the past, it became more difficult to train such networks as the number of layers grew. The two biggest challenges were referred to as the problems of **exploding gradient** and **vanishing gradient** as gradient descent was used to train the network parameters.

While the problem of exploding gradient was easier to deal with by applying simple techniques like **gradient clipping** and L1 or L2 regularization, the problem of vanishing gradient remained intractable for decades.

What is vanishing gradient and why does it arise? To update the values of the parameters in neural networks the algorithm called **backpropagation** is typically used. Backpropagation is an efficient algorithm for computing gradients on neural networks using the chain rule. In Chapter 4, we have already seen how the chain rule is used to calculate partial derivatives of a complex function. During gradient descent, the neural network's parameters receive an update proportional to the partial derivative of the cost function with respect to the current parameter in each iteration of training. The problem is that in some cases, the gradient will be vanishingly small, effectively preventing some parameters from changing their value. In the worst case, this may completely stop the neural network from further training.

Traditional activation functions, such as the hyperbolic tangent function I mentioned above, have gradients in the range $(0, 1)$, and backpropagation computes gradients by the chain rule. That has the effect of multiplying n of these small numbers to compute gradients of the earlier (leftmost) layers in an n-layer network, meaning that the gradient decreases exponentially with n. That results in the effect that the earlier layers train very slowly, if at all.

However, the modern implementations of neural network learning algorithms allow you to effectively train very deep neural networks (up to hundreds of layers). This is due to several improvements combined together, including ReLU, LSTM (and other gated units; we consider them below), as well as techniques such as **skip connections** used in **residual neural networks**, as well as advanced modifications of the gradient descent algorithm.

Therefore, today, since the problems of vanishing and exploding gradient are mostly solved (or their effect diminished) to a great extent, the term "deep learning" refers to training neural networks using the modern algorithmic and mathematical toolkit independently of how deep the neural network is. In practice, many business problems can be solved with neural networks having 2-3 layers between the input and output layers. The layers that are neither input nor output are often called **hidden layers**.

6.2.1 Convolutional Neural Network

You may have noticed that the number of parameters an MLP can have grows very fast as you make your network bigger. More specifically, as you add one layer, you add $(size_{l-1} + 1) \cdot size_l$ parameters (our matrix $\mathbf{W}_l$ plus the vector $\mathbf{b}_l$). That means that if you add another 1000-unit layer to an existing neural network, then you add more than 1 million additional parameters to your model. Optimizing such big models is a very computationally intensive problem.

When our training examples are images, the input is very high-dimensional[3]. If you want to learn to classify images using an MLP, the optimization problem is likely to become intractable.

A **convolutional neural network** (CNN) is a special kind of FFNN that significantly reduces the number of parameters in a deep neural network with many units without losing too much in the quality of the model. CNNs have found applications in image and text processing where they beat many previously established benchmarks.

Because CNNs were invented with image processing in mind, I explain them on the image classification example.

You may have noticed that in images, pixels that are close to one another usually represent the same type of information: sky, water, leaves, fur, bricks, and so on. The exception from the rule are the edges: the parts of an image where two different objects "touch" one another.

If we can train the neural network to recognize regions of the same information as well as the edges, this knowledge would allow the neural network to predict the object represented in the image. For example, if the neural network detected multiple skin regions and edges that look like parts of an oval with skin-like tone on the inside and bluish tone on the outside, then it is likely that it's a face on the sky background. If our goal is to detect people on pictures, the neural network will most likely succeed in predicting a person in this picture.

Having in mind that the most important information in the image is local, we can split the image into square patches using a moving window approach[4]. We can then train multiple smaller regression models at once, each small regression model receiving a square patch as input. The goal of each small regression model is to learn to detect a specific kind of pattern in the input patch. For example, one small regression model will learn to detect the sky; another one will detect the grass, the third one will detect edges of a building, and so on.

In CNNs, a small regression model looks like the one in Figure 6.1, but it only has the layer 1 and doesn't have layers 2 and 3. To detect some pattern, a small regression model has to learn the parameters of a matrix $\mathbf{F}$ (for "filter") of size $p \times p$, where p is the size of a patch. Let's assume, for simplicity, that the input image is black and white, with 1 representing black and 0 representing white pixels. Assume also that our patches are 3 by 3 pixels ($p = 3$). Some patch could then look like the following matrix $\mathbf{P}$ (for "patch"):

$$\mathbf{P} = \begin{bmatrix} 0 & 1 & 0 \\ 1 & 1 & 1 \\ 0 & 1 & 0 \end{bmatrix}.$$

---

[3]Each pixel of an image is a feature. If our image is 100 by 100 pixels, then there are 10,000 features.

[4]Consider this as if you looked at a dollar bill in a microscope. To see the whole bill you have to gradually move your bill from left to right and from top to bottom. At each moment in time, you see only a part of the bill of fixed dimensions. This approach is called **moving window**.

The above patch represents a pattern that looks like a cross. The small regression model that will detect such patterns (and only them) would need to learn a 3 by 3 parameter matrix $\mathbf{F}$ where parameters at positions corresponding to the 1s in the input patch would be positive numbers, while the parameters in positions corresponding to 0s would be close to zero. If we calculate the **convolution** of matrices $\mathbf{P}$ and $\mathbf{F}$, the value we obtain is higher the more similar $\mathbf{F}$ is to $\mathbf{P}$. To illustrate the convolution of two matrices, assume that $\mathbf{F}$ looks like this:

$$\mathbf{F} = \begin{bmatrix} 0 & 2 & 3 \\ 2 & 4 & 1 \\ 0 & 3 & 0 \end{bmatrix}.$$

Then convolution operator is only defined for matrices that have the same number of rows and columns. For our matrices of $\mathbf{P}$ and $\mathbf{F}$ it's calculated as illustrated below:

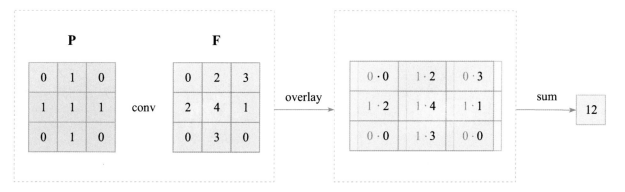

Figure 6.2: A convolution between two matrices.

If our input patch $\mathbf{P}$ had a different patten, for example, that of a letter T,

$$\mathbf{P} = \begin{bmatrix} 1 & 1 & 1 \\ 0 & 1 & 0 \\ 0 & 1 & 0 \end{bmatrix},$$

then the convolution with $\mathbf{F}$ would give a lower result: 9. So, you can see the more the patch "looks" like the filter, the higher the value of the convolution operation is. For convenience, there's also a bias parameter b associated with each filter $\mathbf{F}$ which is added to the result of a convolution before applying the nonlinearity (activation function).

One layer of a CNN consists of multiple convolution filters (each with its own bias parameter), just like one layer in a vanilla FFNN consists of multiple units. Each filter of the first (leftmost) layer slides — or *convolves* — across the input image, left to right, top to bottom, and convolution is computed at each iteration.

An illustration of the process is given in Figure 6.3 where 6 steps of one filter convolving across an image are shown.

The filter matrix (one for each filter in each layer) and bias values are trainable parameters that are optimized using gradient descent with backpropagation.

A nonlinearity is applied to the sum of the convolution and the bias term. Typically, the ReLU activation function is used in all hidden layers. The activation function of the output layer depends on the task.

Since we can have $size_l$ filters in each layer l, the output of the convolution layer l would consist of $size_l$ matrices, one for each filter.

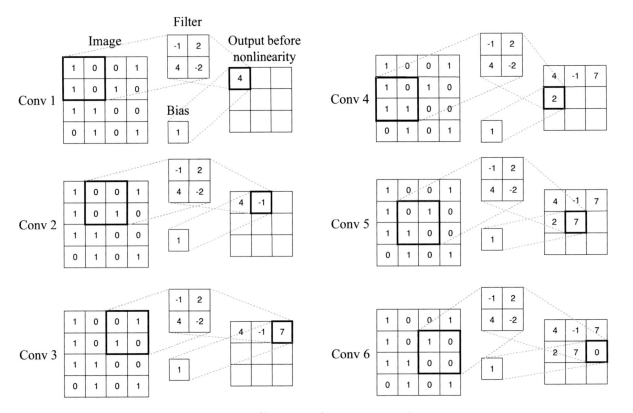

Figure 6.3: A filter convolving across an image.

If the CNN has one convolution layer following another convolution layer, then the subsequent layer $l + 1$ treats the output of the preceding layer l as a collection of $size_l$ image matrices. Such a collection is called a **volume**. The size of that collection is called the volume's depth. Each filter of layer $l + 1$ convolves the whole volume. The convolution of a patch of a volume is simply the sum of convolutions of the corresponding patches of individual matrices the volume consists of.

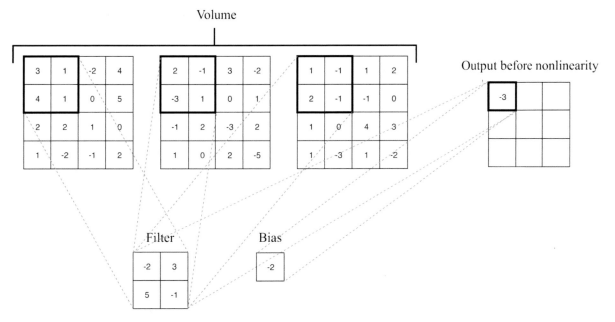

Figure 6.4: Convolution of a volume consisting of three matrices.

An example of a convolution of a patch of a volume consisting of depth 3 is shown in Figure 6.4. The value of the convolution, -3, was obtained as $(-2 \cdot 3 + 3 \cdot 1 + 5 \cdot 4 + -1 \cdot 1) + (-2 \cdot 2 + 3 \cdot (-1) + 5 \cdot (-3) + -1 \cdot 1) + (-2 \cdot 1 + 3 \cdot (-1) + 5 \cdot 2 + -1 \cdot (-1)) + (-2)$.

In computer vision, CNNs often get volumes as input, since an image is usually represented by three channels: R, G, and B, each channel being a monochrome picture.

Two important properties of convolution are **stride** and **padding**. Stride is the step size of the moving window. In Figure 6.3, the stride is 1, that is the filter slides to the right and to the bottom by one cell at a time. In Figure 6.5, you can see a partial example of convolution with stride 2. You can see that the output matrix is smaller when stride is bigger.

Padding allows getting a larger output matrix; it's the width of the square of additional cells with which you surround the image (or volume) before you convolve it with the filter. The cells added by padding usually contain zeroes. In Figure 6.3, the padding is 0, so no additional cells are added to the image. In Figure 6.6, on the other hand, the stride is 2 and padding is 1, so a square of width 1 of additional cells are added to the image. You can see that the output matrix is bigger when padding is bigger[5].

---

[5] To save space, in Figure 6.6, only the first two of the nine convolutions are shown.

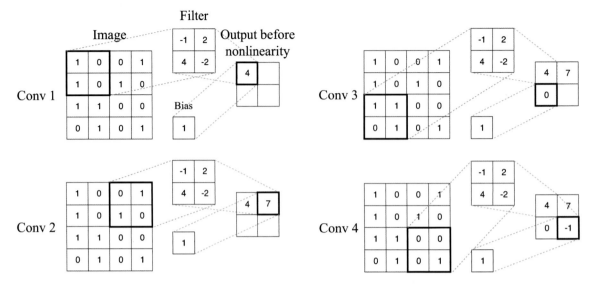

Figure 6.5: Convolution with stride 2.

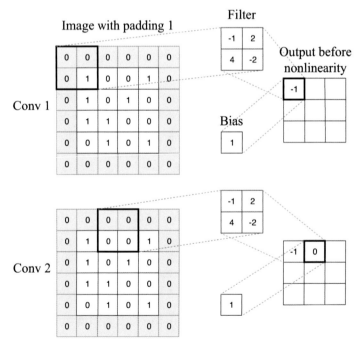

Figure 6.6: Convolution with stride 2 and padding 1.

An example of an image with padding 2 is shown in Figure 6.7. Padding is helpful with larger filters because it allows them to better "scan" the boundaries of the image.

0	0	0	0	0	0	0	0
0	0	0	0	0	0	0	0
0	0	1	0	0	1	0	0
0	0	1	0	1	0	0	0
0	0	1	1	0	0	0	0
0	0	0	1	0	1	0	0
0	0	0	0	0	0	0	0
0	0	0	0	0	0	0	0

Figure 6.7: Image with padding 2.

This section would not be complete without presenting **pooling**, a technique very often used in CNNs. Pooling works in a way very similar to convolution, as a filter applied using a moving window approach. However, instead of applying a trainable filter to an input matrix or a volume, pooling layer applies a fixed operator, usually either max or average. Similarly to convolution, pooling has hyperparameters: the size of the filter and the stride. An example of max pooling with filter of size 2 and stride 2 is shown in Figure 6.8.

Usually, a pooling layer follows a convolution layer, and it gets the output of convolution as input. When pooling is applied to a volume, each matrix in the volume is processed independently of others. Therefore, the output of the pooling layer applied to a volume is a volume of the same depth as the input.

As you can see, pooling only has hyperparameters and doesn't have parameters to learn. Typically, the filter of size 2 or 3 and stride 2 are used in practice. Max pooling is more popular than average and often gives better results.

Typically pooling contributes to the increased accuracy of the model. It also improves the speed of training by reducing the number of parameters of the neural network. (As you can see in Figure 6.8, with filter size 2 and stride 2 the number of parameters is reduced to 25%, that is to 4 parameters instead of 16.)

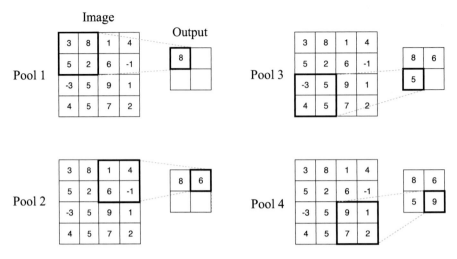

Figure 6.8: Pooling with filter of size 2 and stride 2.

6.2.2 Recurrent Neural Network

Recurrent neural networks (RNNs) are used to label, classify, or generate sequences. A sequence is a matrix, each row of which is a feature vector and the order of rows matters. To label a sequence is to predict a class for each feature vector in a sequence. To classify a sequence is to predict a class for the entire sequence. To generate a sequence is to output another sequence (of a possibly different length) somehow relevant to the input sequence.

RNNs are often used in text processing because sentences and texts are naturally sequences of either words/punctuation marks or sequences of characters. For the same reason, recurrent neural networks are also used in speech processing.

A recurrent neural network is not feed-forward: it contains loops. The idea is that each unit u of recurrent layer l has a real-valued **state** $h_{l,u}$. The state can be seen as the memory of the unit. In RNN, each unit u in each layer l receives two inputs: a vector of states from the previous layer $l - 1$ and the vector of states from this same layer l from *the previous time step*.

To illustrate the idea, let's consider the first and the second recurrent layers of an RNN. The first (leftmost) layer receives a feature vector as input. The second layer receives the output of the first layer as input.

This situation is schematically depicted in Figure 6.9. As I said above, each training example is a matrix in which each row is a feature vector. For simplicity, let's illustrate this matrix as a sequence of vectors $\mathbf{X} = [\mathbf{x}^1, \mathbf{x}^2, \ldots, \mathbf{x}^{t-1}, \mathbf{x}^t, \mathbf{x}^{t+1}, \ldots, \mathbf{x}^{length_\mathbf{X}}]$, where $length_\mathbf{X}$ is the length of the input sequence. If our input example $\mathbf{X}$ is a text sentence, then feature vector $\mathbf{x}^t$ for each $t = 1, \ldots, length_\mathbf{X}$ represents a word in the sentence at position t.

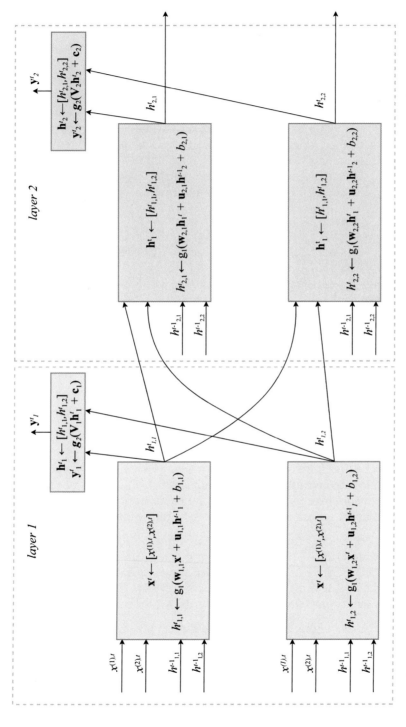

Figure 6.9: The first two layers of an RNN. The input feature vector is two-dimensional; each layer has two units.

As depicted in Figure 6.9, in an RNN, the feature vectors from an input example are "read" by the neural network sequentially in the order of the timesteps. The index t denotes a timestep. To update the state $h_{l,u}^t$ at each timestep t in each unit u of each layer l we first calculate a linear combination of the input feature vector with the state vector $\mathbf{h}_{l,u}^{t-1}$ of this same layer from the previous timestep, $t-1$. The linear combination of two vectors is calculated using two parameter vectors $\mathbf{w}_{l,u}$, $\mathbf{u}_{l,u}$ and a parameter $b_{l,u}$. The value of $h_{l,u}^t$ is then obtained by applying activation function g_1 to the result of the linear combination. A typical choice for function g_1 is $tanh$. The output $\mathbf{y}_l^t$ is typically a vector calculated for the whole layer l at once. To obtain $\mathbf{y}_l^t$, we use activation function g_2 that takes a vector as input and returns a different vector of the same dimensionality. The function g_2 is applied to a linear combination of the state vector values $h_{l,u}^t$ calculated using a parameter matrix $\mathbf{V}_l$ and a parameter vector $\mathbf{c}_{l,u}$. In classification, a typical choice for g_2 is the **softmax function**:

$$\boldsymbol{\sigma}(\mathbf{z}) \overset{\text{def}}{=} [\sigma^{(1)}, \ldots, \sigma^{(D)}], \text{ where } \sigma^{(j)} \overset{\text{def}}{=} \frac{\exp\left(z^{(j)}\right)}{\sum_{k=1}^{D} \exp\left(z^{(k)}\right)}.$$

The softmax function is a generalization of the sigmoid function to multidimensional outputs. It has the property that $\sum_{j=1}^{D} \sigma^{(j)} = 1$ and $\sigma^{(j)} > 0$ for all j.

The dimensionality of $\mathbf{V}_l$ is chosen by the data analyst such that multiplication of matrix $\mathbf{V}_l$ by the vector $\mathbf{h}_l^t$ results in a vector of the same dimensionality as that of the vector $\mathbf{c}_l$. This choice depends on the dimensionality for the output label $\mathbf{y}$ in your training data. (Until now we only saw one-dimensional labels, but we will see in the future chapters that labels can be multidimensional as well.)

The values of $\mathbf{w}_{l,u}$, $\mathbf{u}_{l,u}$, $b_{l,u}$, $\mathbf{V}_{l,u}$, and $\mathbf{c}_{l,u}$ are computed from the training data using gradient descent with backpropagation. To train RNN models, a special version of backpropagation is used called **backpropagation through time.**

Both *tanh* and *softmax* suffer from the vanishing gradient problem. Even if our RNN has just one or two recurrent layers, because of the sequential nature of the input, backpropagation has to "unfold" the network over time. From the point of view of the gradient calculation, in practice this means that the longer is the input sequence, the deeper is the unfolded network.

Another problem RNNs have is that of handling long-term dependencies. As the length of the input sequence grows, the feature vectors from the beginning of the sequence tend to be "forgotten," because the state of each unit, which serves as network's memory, becomes significantly affected by the feature vectors read more recently. Therefore, in text or speech processing, the cause-effect link between distant words in a long sentence can be lost.

The most effective recurrent neural network models used in practice are **gated RNNs**. These include the **long short-term memory** (LSTM) networks and networks based on the **gated recurrent unit** (GRU).

The beauty of using gated units in RNNs is that such networks can store information in their units for future use, much like bits in a computer's memory. The difference with the real memory is that reading, writing, and erasure of information stored in each unit is controlled by activation functions that take values in the range $(0, 1)$. The trained neural network can "read" the input sequence of feature vectors and decide at some early time step t to keep specific information about the feature vectors. That information about the earlier feature vectors can later be used by the model to process the feature vectors from near the end of the input sequence. For example, if the input text starts with the word *she*, a language processing RNN model could decide to store the information about the gender to interpret correctly the word *their* seen later in the sentence.

Units make decisions about what information to store, and when to allow reads, writes, and erasures. Those decisions are learned from data and implemented through the concept of *gates*. There are several architectures of gated units. A simple but effective one is called the **minimal gated GRU** and is composed of a memory cell, and a forget gate.

Let's look at the math of a GRU unit on an example of the first layer of the RNN (the one that takes the sequence of feature vectors as input). A minimal gated GRU unit u in layer l takes two inputs: the vector of the memory cell values from all units in the same layer from the previous timestep, $\mathbf{h}_l^{t-1}$, and a feature vector $\mathbf{x}^t$. It then uses these two vectors as follows (all operations in the below sequence are executed in the unit one after another):

$$\tilde{h}_{l,u}^t \leftarrow g_1(\mathbf{w}_{l,u}\mathbf{x}^t + \mathbf{u}_{l,u}\mathbf{h}_l^{t-1} + b_{l,u}),$$
$$\Gamma_{l,u}^t \leftarrow g_2(\mathbf{m}_{l,u}\mathbf{x}^t + \mathbf{o}_{l,u}\mathbf{h}_l^{t-1} + a_{l,u}),$$
$$h_{l,u}^t \leftarrow \Gamma_{l,u}^t \tilde{h}_l^t + (1 - \Gamma_{l,u}^t)h_l^{t-1},$$
$$\mathbf{h}_l^t \leftarrow [h_{l,1}^t, \ldots, h_{l,size_l}^t]$$
$$\mathbf{y}_l^t \leftarrow g_3(\mathbf{V}_l\mathbf{h}_l^t + \mathbf{c}_{l,u}),$$

where g_1 is the *tanh* activation function, g_2 is called the gate function and is implemented as the sigmoid function taking values in the range $(0, 1)$. If the gate $\Gamma_{l,u}$ is close to 0, then the memory cell keeps its value from the previous time step, h_l^{t-1}. On the other hand, if the gate $\Gamma_{l,u}$ is close to 1, the value of the memory cell is overwritten by a new value $\tilde{h}_{l,u}^t$ (see the third assignment from the top). Just like in standard RNNs, g_3 is usually softmax.

A gated unit takes an input and stores it for some time. This is equivalent to applying the identity function ($f(x) = x$) to the input. Because the derivative of the identity function is constant, when a network with gated units is trained with backpropagation through time, the gradient does not vanish.

Other important extensions to RNNs include **bi-directional RNNs**, RNNs with **attention** and **sequence-to-sequence RNN** models. The latter, in particular, are frequently used to build neural machine translation models and other models for text to text transformations. A generalization of an RNN is a **recursive neural network**.

Chapter 7

Problems and Solutions

7.1 Kernel Regression

We talked about linear regression, but what if our data doesn't have the form of a straight line? Polynomial regression could help. Let's say we have a one-dimensional data $\{(x_i, y_i)\}_{i=1}^N$. We could try to fit a quadratic line $y = w_1 x_i + w_2 x_i^2 + b$ to our data. By defining the mean squared error (MSE) cost function, we could apply gradient descent and find the values of parameters w_1, w_2, and b that minimize this cost function. In one- or two-dimensional space, we can easily see whether the function fits the data. However, if our input is a D-dimensional feature vector, with $D > 3$, finding the right polynomial would be hard.

Kernel regression is a non-parametric method. That means that there are no parameters to learn. The model is based on the data itself (like in kNN). In its simplest form, in kernel regression we look for a model like this:

$$f(x) = \frac{1}{N} \sum_{i=1}^N w_i y_i, \text{ where } w_i = \frac{N k(\frac{x_i - x}{b})}{\sum_{l=1}^N k(\frac{x_l - x}{b})}. \tag{7.1}$$

The function $k(\cdot)$ is called a **kernel**. The kernel plays the role of a similarity function: the values of coefficients w_i are higher when x is similar to x_i and lower when they are dissimilar. Kernels can have different forms, the most frequently used one is the Gaussian kernel:

$$k(z) = \frac{1}{\sqrt{2\pi}} \exp\left(\frac{-z^2}{2}\right).$$

The value b is a hyperparameter that we tune using the validation set (by running the model built with a specific value of b on the validation set examples and calculating the MSE). You can see an illustration of the influence b has on the shape of the regression line in Figure 7.1.

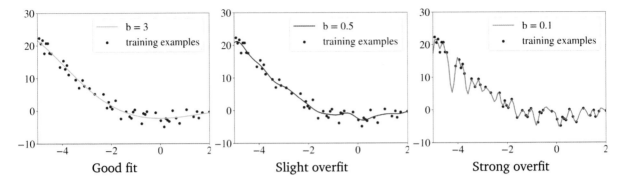

Figure 7.1: Example of kernel regression line with a Gaussian kernel for three values of b.

If your inputs are multi-dimensional feature vectors, the terms $x_i - x$ and $x_l - x$ in eq. 7.1 have to be replaced by Euclidean distance $\|\mathbf{x}_i - \mathbf{x}\|$ and $\|\mathbf{x}_l - \mathbf{x}\|$ respectively.

7.2 Multiclass Classification

Although many classification problems can be defined using two classes, some are defined with more than two classes, which requires adaptations of our machine learning algorithms.

In **multiclass classification**, the label can be one of C classes: $y \in \{1, \ldots, C\}$. Many machine learning algorithms are binary; SVM is an example. Some algorithms can naturally be extended to handle multiclass problems. ID3 and other decision tree learning algorithms can be simply changed like this:

$$f_{ID3}^{S} \stackrel{\text{def}}{=} \Pr(y_i = c|\mathbf{x}) = \frac{1}{|\mathcal{S}|} \sum_{\{y \,|\, (\mathbf{x},y) \in \mathcal{S}, y=c\}} y,$$

for all $c \in \{1, \ldots, C\}$, where $\mathcal{S}$ is the leaf node in which the prediction is made.

Logistic regression can be naturally extended to multiclass learning problems by replacing the sigmoid function with the **softmax function** which we already saw in Chapter 6.

The kNN algorithm is also straightforward to extend to the multiclass case: when we find the k closest examples for the input $\mathbf{x}$ and examine them, we return the class that we saw the most among the k examples.

SVM cannot be naturally extended to multiclass problems. Other algorithms can be implemented more efficiently in the binary case. What should you do if you have a multiclass problem but a binary classification learning algorithm? One common strategy is called **one versus rest**. The idea is to transform a multiclass problem into C binary classification problems and build C binary classifiers. For example, if we have three classes, $y \in \{1, 2, 3\}$, we create copies of the original datasets and modify them. In the first copy, we replace all labels not equal to 1 by 0. In the second copy, we replace all labels not equal to 2 by 0. In the third copy, we replace all labels not equal to 3 by 0. Now we have three binary classification problems where we have to learn to distinguish between labels 1 and 0, 2 and 0, and 3 and 0.

Once we have the three models, to classify the new input feature vector $\mathbf{x}$, we apply the three models to the input, and we get three predictions. We then pick the prediction of a non-zero class which is *the most certain*. Remember that in logistic regression, the model returns not a label but a score (between 0 and 1) that can be interpreted as the probability that the label is positive. We can also interpret this score as the certainty of prediction. In SVM, the analog of certainty is the distance d from the input $\mathbf{x}$ to the decision boundary given by,

$$d \overset{\text{def}}{=} \frac{\mathbf{w}^*\mathbf{x} + b^*}{\|w\|}.$$

The larger the distance, the more certain is the prediction. Most learning algorithm either can be naturally converted to a multiclass case, or they return a score we can use in the one versus rest strategy.

7.3 One-Class Classification

Sometimes we only have examples of one class and we want to train a model that would distinguish examples of this class from everything else.

One-class classification, also known as **unary classification** or **class modeling**, tries to identify objects of a specific class among all objects, by learning from a training set containing only the objects of that class. That is different from and more difficult than the traditional classification problem, which tries to distinguish between two or more classes with the training set containing objects from all classes. A typical one-class classification problem is the classification of the traffic in a secure computer network as normal. In this scenario, there are few, if any, examples of the traffic under an attack or during an intrusion. However, the examples of normal traffic are often in abundance. One-class classification learning algorithms are used for outlier detection, anomaly detection, and novelty detection.

There are several one-class learning algorithms. The most widely used in practice are **one-class Gaussian, one-class k-means, one-class kNN**, and **one-class SVM**.

The idea behind the one-class gaussian is that we model our data as if it came from a Gaussian distribution, more precisely **multivariate normal distribution** (MND). The probability density function (pdf) for MND is given by the following equation:

$$f_{\boldsymbol{\mu},\boldsymbol{\Sigma}}(\mathbf{x}) = \frac{\exp\left(-\frac{1}{2}(\mathbf{x} - \boldsymbol{\mu})^\top \boldsymbol{\Sigma}^{-1}(\mathbf{x} - \boldsymbol{\mu})\right)}{\sqrt{(2\pi)^D |\boldsymbol{\Sigma}|}},$$

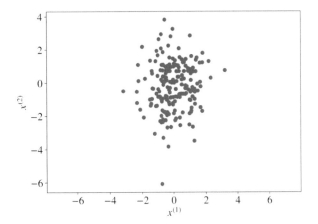
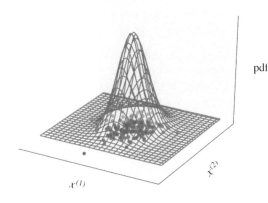

Figure 7.2: One-class classification solved using the one-class gaussian method. Left: two-dimensional feature vectors. Right: the MND curve that maximizes the likelihood of the examples on the left.

where $f_{\mu, \Sigma}(\mathbf{x})$ returns the probability density corresponding to the input feature vector $\mathbf{x}$. Probability density can be interpreted as the likelihood that example $\mathbf{x}$ was drawn from the probability distribution we model as an MND. Values μ (a vector) and Σ (a matrix) are the parameters we have to learn. The **maximum likelihood** criterion (similarly to how we solved the logistic regression learning problem) is optimized to find the optimal values for these two parameters. $|\Sigma| \stackrel{\text{def}}{=} \det \Sigma$ is the *determinant* of the matrix Σ; the notation Σ^{-1} means the *inverse* of the matrix Σ.

If the terms *determinant* and *inverse* are new to you, don't worry. These are standard operations on vector and matrices from the branch of mathematics called *matrix theory*. If you feel the need to know what they are, Wikipedia explains these concepts well.

In practice, the numbers in the vector μ determine the place where the curve of our Gaussian distribution is centered, while the numbers in Σ determine the shape of the curve. For a training set consisting of two-dimensional feature vectors, an example of the one-class Gaussian model is given in Figure 7.2.

Once we have our model parametrized by μ and Σ learned from data, we predict the likelihood of every input $\mathbf{x}$ by using $f_{\mu, \Sigma}(\mathbf{x})$. Only if the likelihood is above a certain threshold, we predict that the example belongs to our class; otherwise, it is classified as the outlier. The value of the threshold is found experimentally or using an "educated guess."

When the data has a more complex shape, a more advanced algorithm can use a combination of several Gaussians (called a mixture of Gaussians). In this case, there are more parameters to learn from data: one μ and one Σ for each Gaussian as well as the parameters that allow combining multiple Gaussians to form one pdf. In Chapter 9, we consider a mixture of

Gaussians with an application to clustering.

One-class k-means and one-class kNN are based on a similar principle as that of one-class Gaussian: build some model of the data and then define a threshold to decide whether our new feature vector looks similar to other examples according to the model. In the former, all training examples are clustered using the **k-means** clustering algorithm and, when a new example x is observed, the distance $d(\mathbf{x})$ is calculated as the minimum distance between x and the center of each cluster. If $d(\mathbf{x})$ is less than a particular threshold, then x belongs to the class.

One-class SVM, depending on formulation, tries either 1) to separate all training examples from the origin (in the feature space) and maximize the distance from the hyperplane to the origin, or 2) to obtain a spherical boundary around the data by minimizing the volume of this hypersphere. I leave the description of the one-class kNN algorithm, as well as the details of the one-class k-means and one-class SVM for the complementary reading.

7.4 Multi-Label Classification

In some situations, more than one label is appropriate to describe an example from the dataset. In this case, we talk about the **multi-label classification**.

For instance, if we want to describe an image, we could assign several labels to it: "conifer," "mountain," "road," all three at the same time (Figure 7.3).

Figure 7.3: A picture labeled as "conifer," "mountain," and "road." Photo: Cate Lagadia.

If the number of possible values for labels is high, but they are all of the same nature, like tags, we can transform each labeled example into several labeled examples, one per label. These new examples all have the same feature vector and only one label. That becomes a multiclass classification problem. We can solve it using the one versus rest strategy. The only difference with the usual multiclass problem is that now we have a new hyperparameter: threshold. If the prediction score for some label is above the threshold, this label is predicted for the input feature vector. In this scenario, multiple labels can be predicted for one feature vector. The value of the threshold is chosen using the validation set.

Analogously, algorithms that naturally can be made multiclass (decision trees, logistic regression and neural networks among others) can be applied to multi-label classification problems. Because they return the score for each class, we can define a threshold and then assign multiple labels to one feature vector if the threshold is above some value.

Neural networks algorithms can naturally train multi-label classification models by using the **binary cross-entropy** cost function. The output layer of the neural network, in this case, has one unit per label. Each unit of the output layer has the sigmoid activation function. Accordingly, each label l is binary ($y_{i,l} \in \{0,1\}$), where $l = 1, \ldots, L$ and $i = 1, \ldots, N$. The binary cross-entropy of predicting the probability $\hat{y}_{i,l}$ that example $\mathbf{x}_i$ has label l is defined as,

$$-(y_{i,l} \ln(\hat{y}_{i,l}) + (1 - y_{i,l}) \ln(1 - \hat{y}_{i,l})).$$

The minimization criterion is simply the average of all binary cross-entropy terms across all training examples and all labels of those examples.

In cases where the number of possible values each label can take is small, one can convert multilabel into a multiclass problem using a different approach. Imagine the following problem. We want to label images and labels can be of two types. The first type of label can have two possible values: $\{photo, painting\}$; the label of the second type can have three possible values $\{portrait, paysage, other\}$. We can create a new fake class for each combination of the two original classes, like this:

Fake Class	Real Class 1	Real Class 2
1	photo	portrait
2	photo	paysage
3	photo	other
4	painting	portrait
5	painting	paysage
6	painting	other

Now we have the same labeled examples, but we replace real multi-labels with one fake label with values from 1 to 6. This approach works well in practice when there are not too many possible combinations of classes. Otherwise, you need to use much more training data to compensate for an increased set of classes.

The primary advantage of this latter approach is that you keep your labels correlated, contrary to the previously seen methods that predict each label independently of one another. Correlation between labels can be essential in many problems. For example, if you want to predict whether an email is *spam* or *not_spam* at the same time as predicting whether it's *ordinary* or *priority* email. You would like to avoid predictions like $[spam, priority]$.

7.5   Ensemble Learning

The fundamental algorithms that we considered in Chapter 3 have their limitations. Because of their simplicity, sometimes they cannot produce a model accurate enough for your problem. You could try using deep neural networks. However, in practice, deep neural networks require a significant amount of labeled data which you might not have. Another approach to boost the performance of simple learning algorithms is **ensemble learning**.

Ensemble learning is a learning paradigm that, instead of trying to learn one super-accurate model, focuses on training a large number of low-accuracy models and then combining the predictions given by those *weak* models to obtain a high-accuracy **meta-model**.

Low-accuracy models are usually learned by **weak learners**, that is, learning algorithms that cannot learn complex models, and thus are typically fast at the training and at the prediction time. The most frequently used weak learner is a decision tree learning algorithm in which we often stop splitting the training set after just a few iterations. The obtained trees are shallow and not particularly accurate, but the idea behind ensemble learning is that if the trees are not identical and each tree is at least slightly better than random guessing, then we can obtain high accuracy by combining a large number of such trees.

To obtain the prediction for input $\mathbf{x}$, the predictions of each weak model are combined using some sort of weighted voting. The specific form of vote weighting depends on the algorithm, but, independently of the algorithm, the idea is the same: if the council of weak models predicts that the message is spam, then we assign the label *spam* to $\mathbf{x}$.

Two principal ensemble learning methods are **boosting** and **bagging**.

7.5.1   Boosting and Bagging

Boosting consists of using the original training data and iteratively creating multiple models by using a weak learner. Each new model would be different from the previous ones in the sense that the weak learner, by building each new model tries to "fix" the errors which previous models make. The final **ensemble model** is a certain combination of those multiple weak models built iteratively.

Bagging consists of creating many "copies" of the training data (each copy is slightly different from another) and then apply the weak learner to each copy to obtain multiple weak models and then combine them. A widely used and effective machine learning algorithm based on the idea of bagging is **random forest**.

7.5.2 Random Forest

The "vanilla" bagging algorithm works as follows. Given a training set, we create B random samples $\mathcal{S}_b$ (for each $b = 1, \ldots, B$) of the training set and build a decision tree model f_b using each sample $\mathcal{S}_b$ as the training set. To sample $\mathcal{S}_b$ for some b, we do the **sampling with replacement**. This means that we start with an empty set, and then pick at random an example from the training set and put its exact copy to $\mathcal{S}_b$ by keeping the original example in the original training set. We keep picking examples at random until the $|\mathcal{S}_b| = N$.

After training, we have B decision trees. The prediction for a new example $\mathbf{x}$ is obtained as the average of B predictions:

$$y \leftarrow \hat{f}(\mathbf{x}) \stackrel{\text{def}}{=} \frac{1}{B} \sum_{b=1}^{B} f_b(\mathbf{x}),$$

in the case of regression, or by taking the majority vote in the case of classification.

Random forest is different from the vanilla bagging in just one way. It uses a modified tree learning algorithm that inspects, at each split in the learning process, a random subset of the features. The reason for doing this is to avoid the correlation of the trees: if one or a few features are very strong predictors for the target, these features will be selected to split examples in many trees. This would result in many correlated trees in our "forest." Correlated predictors cannot help in improving the accuracy of prediction. The main reason behind a better performance of model ensembling is that models that are good will likely agree on the same prediction, while bad models will likely disagree on different ones. Correlation will make bad models more likely to agree, which will hamper the majority vote or the average.

The most important hyperparameters to tune are the number of trees, B, and the size of the random subset of the features to consider at each split.

Random forest is one of the most widely used ensemble learning algorithms. Why is it so effective? The reason is that by using multiple samples of the original dataset, we reduce the **variance** of the final model. Remember that the low variance means low **overfitting**. Overfitting happens when our model tries to explain small variations in the dataset because our dataset is just a small sample of the population of all possible examples of the phenomenon we try to model. If we were unlucky with how our training set was sampled, then it could contain some undesirable (but unavoidable) artifacts: noise, outliers and over- or underrepresented examples. By creating multiple random samples with replacement of our training set, we reduce the effect of these artifacts.

7.5.3 Gradient Boosting

Another effective ensemble learning algorithm, based on the idea of boosting, is **gradient boosting**. Let's first look at gradient boosting for regression. To build a strong regressor, we start with a constant model $f = f_0$ (just like we did in ID3):

$$f = f_0(\mathbf{x}) \overset{\text{def}}{=} \frac{1}{N} \sum_{i=1}^{N} y_i.$$

Then we modify labels of each example $i = 1, \ldots, N$ in our training set as follows:

$$\hat{y}_i \leftarrow y_i - f(\mathbf{x}_i), \tag{7.2}$$

where $\hat{y}_i$, called the **residual**, is the new label for example $\mathbf{x}_i$.

Now we use the modified training set, with residuals instead of original labels, to build a new decision tree model, f_1. The boosting model is now defined as $f \overset{\text{def}}{=} f_0 + \alpha f_1$, where α is the learning rate (a hyperparameter).

Then we recompute the residuals using eq. 7.2 and replace the labels in the training data once again, train the new decision tree model f_2, redefine the boosting model as $f \overset{\text{def}}{=} f_0 + \alpha f_1 + \alpha f_2$ and the process continues until the predefined maximum M of trees are combined.

Intuitively, what's happening here? By computing the residuals, we find how well (or poorly) the target of each training example is predicted by the current model f. We then train another tree to fix the errors of the current model (this is why we use residuals instead of real labels) and add this new tree to the existing model with some weight α. Therefore, each additional tree added to the model partially fixes the errors made by the previous trees until the maximum number M (another hyperparameter) of trees are combined.

Now you should reasonably ask why the algorithm is called *gradient* boosting? In gradient boosting, we don't calculate any gradient contrary to what we did in Chapter 4 for linear regression. To see the similarity between gradient boosting and gradient descent remember why we calculated the gradient in linear regression: we did that to get an idea on where we should move the values of our parameters so that the MSE cost function reaches its minimum. The gradient showed the direction, but we didn't know how far we should go in this direction, so we used a small step α at each iteration and then reevaluated our direction. The same happens in gradient boosting. However, instead of getting the gradient directly, we use its proxy in the form of residuals: they show us how the model has to be adjusted so that the error (the residual) is reduced.

The three principal hyperparameters to tune in gradient boosting are the number of trees, the learning rate, and the depth of trees — all three affect model accuracy. The depth of trees also affects the speed of training and prediction: the shorter, the faster.

It can be shown that training on residuals optimizes the overall model f for the mean squared error criterion. You can see the difference with bagging here: boosting reduces the bias (or underfitting) instead of the variance. As such, boosting can overfit. However, by tuning the depth and the number of trees, overfitting can be largely avoided.

Gradient boosting for classification is similar, but the steps are slightly different. Let's consider the binary case. Assume we have M regression decision trees. Similarly to logistic regression, the prediction of the ensemble of decision trees is modeled using the sigmoid function:

$$\Pr(y = 1|\mathbf{x}, f) \overset{\text{def}}{=} \frac{1}{1 + e^{-f(\mathbf{x})}},$$

where $f(\mathbf{x}) \overset{\text{def}}{=} \sum_{m=1}^{M} f_m(\mathbf{x})$ and f_m is a regression tree.

Again, like in logistic regression, we apply the maximum likelihood principle by trying to find such an f that maximizes $L_f = \sum_{i=1}^{N} \ln\left[\Pr(y_i = 1|\mathbf{x}_i, f)\right]$. Again, to avoid numerical overflow, we maximize the sum of log-likelihoods rather than the product of likelihoods.

The algorithm starts with the initial constant model $f = f_0 = \frac{p}{1-p}$, where $p = \frac{1}{N}\sum_{i=1}^{N} y_i$. (It can be shown that such initialization is optimal for the sigmoid function.) Then at each iteration m, a new tree f_m is added to the model. To find the best f_m, first the partial derivative g_i of the current model is calculated for each $i = 1, \ldots, N$:

$$g_i = \frac{dL_f}{df},$$

where f is the ensemble classifier model built at the previous iteration $m - 1$. To calculate g_i we need to find the derivatives of $\ln\left[\Pr(y_i = 1|\mathbf{x}_i, f)\right]$ with respect to f for all i. Notice that $\ln\left[\Pr(y_i = 1|\mathbf{x}_i, f)\right] \overset{\text{def}}{=} \ln\left[\frac{1}{1 + e^{-f(\mathbf{x}_i)}}\right]$. The derivative of the right-hand term in the previous equation with respect to f equals $\frac{1}{e^{f(\mathbf{x}_i)} + 1}$.

We then transform our training set by replacing the original label y_i with the corresponding partial derivative g_i, and build a new tree f_m using the transformed training set. Then we find the optimal update step ρ_m as:

$$\rho_m \leftarrow \arg\max_{\rho} L_{f + \rho f_m}.$$

At the end of iteration m, we update the ensemble model f by adding the new tree f_m:

$$f \leftarrow f + \alpha\rho_m f_m.$$

We iterate until $m = M$, then we stop and return the ensemble model f.

Gradient boosting is one of the most powerful machine learning algorithms—not just because it creates very accurate models, but also because it is capable of handling huge datasets with millions of examples and features. It usually outperforms random forest in accuracy but, because of its sequential nature, can be significantly slower in training.

7.6 Learning to Label Sequences

Sequence is one the most frequently observed types of structured data. We communicate using sequences of words and sentences, we execute tasks in sequences, our genes, the music we listen and videos we watch, our observations of a continuous process, such as a moving car or the price of a stock are all sequential.

Sequence labeling is the problem of automatically assigning a label to each element of a sequence. A labeled sequential training example in sequence labeling is a pair of lists $(\mathbf{X}, \mathbf{Y})$, where $\mathbf{X}$ is a list of feature vectors, one per time step, $\mathbf{Y}$ is a list of the same length of labels. For example, $\mathbf{X}$ could represent words in a sentence such as ["big", "beautiful", "car"], and $\mathbf{Y}$ would be the list of the corresponding parts of speech, such as ["adjective", "adjective", "noun"]). More formally, in an example i, $\mathbf{X}_i = [\mathbf{x}_i^1, \mathbf{x}_i^2, \ldots, \mathbf{x}_i^{size_i}]$, where $size_i$ is the length of the sequence of the example i, $\mathbf{Y}_i = [y_i^1, y_i^2, \ldots, y_i^{size_i}]$ and $y_i \in \{1, 2, \ldots, C\}$.

You have already seen that an RNN can be used to label a sequence. At each time step t, it reads an input feature vector $\mathbf{x}_i^{(t)}$, and the last recurrent layer outputs a label $y_{last}^{(t)}$ (in the case of binary labeling) or $\mathbf{y}_{last}^{(t)}$ (in the case of multiclass or multilabel labeling).

However, RNN is not the only possible model for sequence labeling. The model called **Conditional Random Fields** (CRF) is a very effective alternative that often performs well in practice for the feature vectors that have many informative features. For example, imagine we have the task of **named entity extraction** and we want to build a model that would label each word in the sentence such as "I go to San Francisco" with one of the following classes: $\{location, name, company\_name, other\}$. If our feature vectors (which represent words) contain such binary features as "whether or not the word starts with a capital letter" and "whether or not the word can be found in the list of locations," such features would be very informative and help to classify the words *San* and *Francisco* as *location*.

Building handcrafted features is known to be a labor-intensive process that requires a significant level of domain expertise.

CRF is an interesting model and can be seen as a generalization of logistic regression to sequences. However, in practice, for sequence labeling tasks, it has been outperformed by bidirectional deep gated RNN. CRFs are also significantly slower in training which makes them difficult to apply to large training sets (with hundreds of thousands of examples). Additionally, a large training set is where a deep neural network thrives.

7.7 Sequence-to-Sequence Learning

Sequence-to-sequence learning (often abbreviated as seq2seq learning) is a generalization of the sequence labeling problem. In seq2seq, X_i and Y_i can have different lengths. seq2seq models have found application in machine translation (where, for example, the input is an English sentence, and the output is the corresponding French sentence), conversational interfaces (where the input is a question typed by the user, and the output is the answer from the machine), text summarization, spelling correction, and many others.

Many but not all seq2seq learning problems are currently best solved by neural networks. The network architectures used in seq2seq all have two parts: an **encoder** and a **decoder**.

In seq2seq neural network learning, the encoder is a neural network that accepts sequential input. It can be an RNN, but also a CNN or some other architecture. The role of the encoder is to read the input and generate some sort of state (similar to the state in RNN) that can be seen as a numerical representation of the *meaning* of the input the machine can work with. The meaning of some entity, whether it be an image, a text or a video, is usually a vector or a matrix that contains real numbers. In machine learning jargon, this vector (or matrix) is called the **embedding** of the input.

The decoder is another neural network that takes an embedding as input and is capable of generating a sequence of outputs. As you could have already guessed, that embedding comes from the encoder. To produce a sequence of outputs, the decoder takes a *start of sequence* input feature vector $\mathbf{x}^{(0)}$ (typically all zeroes), produces the first output $\mathbf{y}^{(1)}$, updates its state by combining the embedding and the input $\mathbf{x}^{(0)}$, and then uses the output $\mathbf{y}^{(1)}$ as its next input $\mathbf{x}^{(1)}$. For simplicity, the dimensionality of $\mathbf{y}^{(t)}$ can be the same as that of $\mathbf{x}^{(t)}$; however, it is not strictly necessary. As we saw in Chapter 6, each layer of an RNN can produce many simultaneous outputs: one can be used to generate the label $\mathbf{y}^{(t)}$, while another one, of different dimensionality, can be used as the $\mathbf{x}^{(t)}$.

Both encoder and decoder are trained simultaneously using the training data. The errors at the decoder output are propagated to the encoder via backpropagation.

A traditional seq2seq architecture is illustrated in Figure 7.4. More accurate predictions can be obtained using an architecture with **attention**. Attention mechanism is implemented by an additional set of parameters that combine some information from the encoder (in RNNs, this information is the list of state vectors of the last recurrent layer from all encoder time steps) and the current state of the decoder to generate the label. That allows for even better retention of long-term dependencies than provided by gated units and bidirectional RNN.

A seq2seq architecture with attention is illustrated in Figure 7.5.

seq2seq learning is a relatively new research domain. Novel network architectures are regularly discovered and published. Training such architectures can be challenging as the

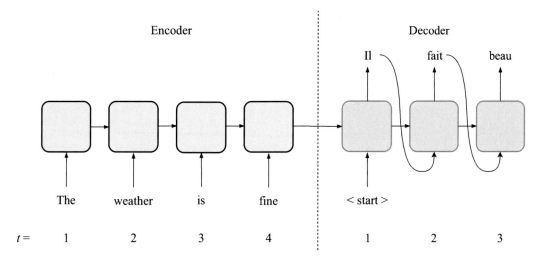

Figure 7.4: A traditional seq2seq architecture. The embedding, usually given by the state of the last layer of the encoder, is passed from the blue to the purple subnetwork.

number of hyperparameters to tune and other architectural decisions can be overwhelming. Consult the book's wiki for the state of the art material, tutorials and code samples.

7.8 Active Learning

Active learning is an interesting supervised learning paradigm. It is usually applied when obtaining labeled examples is costly. That is often the case in the medical or financial domains, where the opinion of an expert may be required to annotate patients' or customers' data. The idea is to start learning with relatively few labeled examples, and a large number of unlabeled ones, and then label only those examples that contribute the most to the model quality.

There are multiple strategies of active learning. Here, we discuss only the following two:

1) data density and uncertainty based, and
2) support vector-based.

The former strategy applies the current model f, trained using the existing labeled examples, to each of the remaining unlabelled examples (or, to save computing time, to some random sample of them). For each unlabeled example $\mathbf{x}$, the following importance score is computed: $density(\mathbf{x}) \cdot uncertainty_f(\mathbf{x})$. Density reflects how many examples surround $\mathbf{x}$ in its close neighborhood, while $uncertainty_f(\mathbf{x})$ reflects how uncertain the prediction of the model f is for $\mathbf{x}$. In binary classification with sigmoid, the closer the prediction score is to 0.5, the more

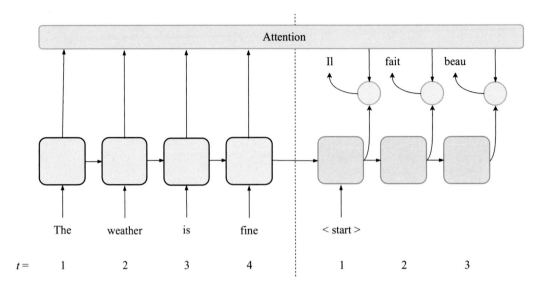

Figure 7.5: A seq2seq architecture with attention.

uncertain is the prediction. In SVM, the closer the example is to the decision boundary, the most uncertain is the prediction.

In multiclass classification, **entropy** can be used as a typical measure of uncertainty:

$$H_f(\mathbf{x}) = -\sum_{c=1}^{C} \Pr(y^{(c)}; f(\mathbf{x})) \ln \left[\Pr(y^{(c)}; f(\mathbf{x})) \right],$$

where $\Pr(y^{(c)}; f(\mathbf{x}))$ is the probability score the model f assigns to class $y^{(c)}$ when classifying $\mathbf{x}$. You can see that if for each $y^{(c)}$, $f(y^{(c)}) = \frac{1}{C}$ then the model is the most uncertain and the entropy is at its maximum of 1; on the other hand, if for some $y^{(c)}$, $f(y^{(c)}) = 1$, then the model is certain about the class $y^{(c)}$ and the entropy is at its minimum of 0.

Density for the example $\mathbf{x}$ can be obtained by taking the average of the distance from $\mathbf{x}$ to each of its k nearest neighbors (with k being a hyperparameter).

Once we know the importance score of each unlabeled example, we pick the one with the highest importance score and ask the expert to annotate it. Then we add the new annotated example to the training set, rebuild the model and continue the process until some stopping criterion is satisfied. A stopping criterion can be chosen in advance (the maximum number of requests to the expert based on the available budget) or depend on how well our model performs according to some metric.

The support vector-based active learning strategy consists in building an SVM model using the labeled data. We then ask our expert to annotate the unlabeled example that lies the closest to the hyperplane that separates the two classes. The idea is that if the example lies closest to the hyperplane, then it is the least certain and would contribute the most to the reduction of possible places where the true (the one we look for) hyperplane could lie.

Some active learning strategies can incorporate the cost of asking an expert for a label. Others *learn* to ask expert's opinion. The "query by committee" strategy consists of training multiple models using different methods and then asking an expert to label example on which those models disagree the most. Some strategies try to select examples to label so that the variance or the bias of the model are reduced the most.

7.9 Semi-Supervised Learning

In **semi-supervised learning** (SSL) we also have labeled a small fraction of the dataset; most of the remaining examples are unlabeled. Our goal is to leverage a large number of unlabeled examples to improve the model performance without asking for additional labeled examples.

Historically, there were multiple attempts at solving this problem. None of them could be called universally acclaimed and frequently used in practice. For example, one frequently cited SSL method is called **self-learning**. In self-learning, we use a learning algorithm to build the initial model using the labeled examples. Then we apply the model to all unlabeled examples and label them using the model. If the confidence score of prediction for some unlabeled example x is higher than some threshold (chosen experimentally), then we add this labeled example to our training set, retrain the model and continue like this until a stopping criterion is satisfied. We could stop, for example, if the accuracy of the model has not been improved during the last m iterations.

The above method can bring some improvement to the model compared to just using the initially labeled dataset, but the increase in performance usually is not impressive. Furthermore, in practice, the quality of the model could even decrease. That depends on the properties of the statistical distribution the data was drawn from, which is usually unknown.

On the other hand, the recent advancements in neural network learning brought some impressive results. For example, it was shown that for some datasets, such as MNIST (a frequent testbench in computer vision that consists of labeled images of handwritten digits from 0 to 9) the model trained in a semi-supervised way has an almost perfect performance with just 10 labeled examples per class (100 labeled examples overall). For comparison, MNIST contains 70,000 labeled examples (60,000 for training and 10,000 for test). The neural network architecture that attained such a remarkable performance is called a **ladder network**. To understand ladder networks you have to understand what an **autoencoder** is.

An autoencoder is a feed-forward neural network with an encoder-decoder architecture. It is trained to reconstruct its input. So the training example is a pair $(\mathbf{x}, \mathbf{x})$. We want the output $\hat{\mathbf{x}}$ of the model $f(\mathbf{x})$ to be as similar to the input $\mathbf{x}$ as possible.

An important detail here is that an autoencoder's network looks like an hourglass with a **bottleneck layer** in the middle that contains the embedding of the D-dimensional input vector; the embedding layer usually has much fewer units than D. The goal of the decoder is to reconstruct the input feature vector from this embedding. Theoretically, it is sufficient to have 10 units in the bottleneck layer to successfully encode MNIST images. In a typical autoencoder schematically depicted in Figure 7.6, the cost function is usually either the mean squared error (when features can be any number) or the binary cross-entropy (when features are binary and the units of the last layer of the decoder have the sigmoid activation function). If the cost is the mean squared error, then it is given by:

$$\frac{1}{N}\sum_{i=1}^{N}\|\mathbf{x}_i - f(\mathbf{x}_i)\|^2,$$

where $\|\mathbf{x}_i - f(\mathbf{x}_i)\|$ is the Euclidean distance between two vectors.

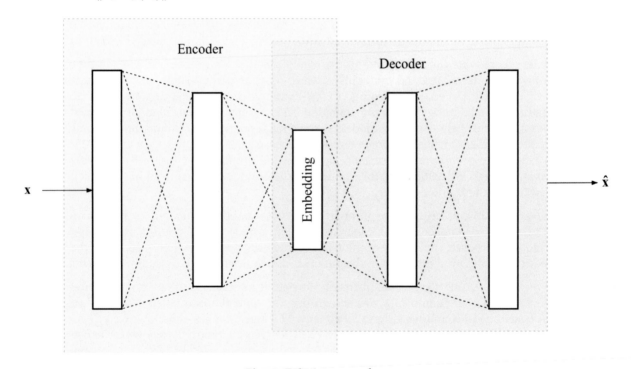

Figure 7.6: Autoencoder.

A **denoising autoencoder** corrupts the left-hand side $\mathbf{x}$ in the training example $(\mathbf{x}, \mathbf{x})$ by adding some random perturbation to the features. If our examples are grayscale images with pixels represented as values between 0 and 1, usually a **Gaussian noise** is added to each feature. For each feature j of the input feature vector $\mathbf{x}$ the noise value $n^{(j)}$ is sampled from the **Gaussian distribution**:

$$n^{(j)} \sim \mathcal{N}(\mu, \sigma^2),$$

where the notation $\sim$ means "sampled from," and $\mathcal{N}(\mu, \sigma^2)$ denotes the Gaussian distribution with mean μ and standard deviation σ whose pdf is given by:

$$f_{\boldsymbol{\theta}}(z) = \frac{1}{\sigma\sqrt{2\pi}} \exp\left(-\frac{(z-\mu)^2}{2\sigma^2}\right).$$

In the above equation, π is the constant $3.14159\ldots$ and $\boldsymbol{\theta} \stackrel{\text{def}}{=} [\mu, \sigma]$ is a hyperparameter. The new, corrupted value of the feature $x^{(j)}$ is given by $x^{(j)} + n^{(j)}$.

A **ladder network** is a denoising autoencoder with an upgrade. The encoder and the decoder have the same number of layers. The bottleneck layer is used directly to predict the label (using the softmax activation function). The network has several cost functions. For each layer l of the encoder and the corresponding layer l of the decoder, one cost C_d^l penalizes the difference between the outputs of the two layers (using the squared Euclidean distance). When a labeled example is used during training, another cost function, C_c, penalizes the error in prediction of the label (the negative log-likelihood cost function is used). The combined cost function, $C_c + \sum_{l=1}^{L} \lambda_l C_d^l$ (averaged over all examples in the batch), is optimized by the minibatch stochastic gradient descent with backpropagation. The hyperparameters λ_l for each layer l determine the tradeoff between the classification and encoding-decoding cost.

In the ladder network, not just the input is corrupted with the noise, but also the output of each encoder layer (during training). When we apply the trained model to the new input $\mathbf{x}$ to predict its label, we do not corrupt the input.

Other semi-supervised learning techniques, not related to training neural networks, exist. One of them implies building the model using the labeled data and then clustering the unlabeled and labeled examples together using any clustering technique (we consider some of them in Chapter 9). For each new example, we then output as a prediction the majority label in the cluster it belongs to.

Another technique, called S3VM, is based on using SVM. We build one SVM model for each possible labeling of unlabeled examples and then we pick the model with the largest margin. The paper on S3VM describes an approach that allows solving this problem without actually enumerating all possible labelings.

7.10 One-Shot Learning

This chapter would be incomplete without mentioning two other important supervised learning paradigms. One of them is **one-shot learning**. In one-shot learning, typically applied in face recognition, we want to build a model that can recognize that two photos of the same person represent that same person. If we present to the model two photos of two different people, we expect the model to recognize that the two people are different.

To solve such a problem, we could go a traditional way and build a binary classifier that takes two images as input and predicts either true (when the two pictures represent the same person) or false (when the two pictures belong to different people). However, in practice, this would result in a neural network twice as big as a typical neural network, because each of the two pictures needs its own embedding subnetwork. Training such a network would be challenging not only because of its size but also because the positive examples would be much harder to obtain than negative ones. So the problem is highly imbalanced.

One way to effectively solve the problem is to train a **siamese neural network** (SNN). An SNN can be implemented as any kind of neural network, a CNN, an RNN, or an MLP. The network only takes one image as input at a time; so the size of the network is not doubled. To obtain a binary classifier "same_person"/"not_same" out of a network that only takes one picture as input, we train the networks in a special way.

To train an SNN, we use the **triplet loss** function. For example, let us have three images of a face: image A (for anchor), image P (for positive) and image N (for negative). A and P are two different pictures of the same person; N is a picture of another person. Each training example i is now a triplet (A_i, P_i, N_i).

Let's say we have a neural network model f that can take a picture of a face as input and output an embedding of this picture. The triplet loss for example i is defined as,

$$\max(\|f(A_i) - f(P_i)\|^2 - \|f(A_i) - f(N_i)\|^2 + \alpha, 0). \tag{7.3}$$

The cost function is defined as the average triplet loss:

$$\frac{1}{N} \sum_{i=1}^{N} \max(\|f(A_i) - f(P_i)\|^2 - \|f(A_i) - f(N_i)\|^2 + \alpha, 0),$$

where α is a positive hyperparameter. Intuitively, $\|f(A) - f(P)\|^2$ is low when our neural network outputs similar embedding vectors for A and P; $\|f(A_i) - f(N_i)\|^2$ is high when the embedding for pictures of two different people are different. If our model works the way we want, then the term $m = \|f(A_i) - f(P_i)\|^2 - \|f(A_i) - f(N_i)\|^2$ will always be negative, because we subtract a high value from a small value. By setting α higher, we force the term m to be even smaller, to make sure that the model learned to recognize the two same faces and two different faces with a high margin. If m is not small enough, then because of α the cost will be positive, and the model parameters will be adjusted in backpropagation.

Rather than randomly choosing an image for N, a better way to create triplets for training is to use the current model after several epochs of learning and find candidates for N that are similar to A and P according to that model. Using random examples as N would significantly slow down the training because the neural network will easily see the difference between pictures of two random people, so the average triplet loss will be low most of the time and the parameters will not be updated fast enough.

To build an SNN, we first decide on the architecture of our neural network. For example, CNN is a typical choice if our inputs are images. Given an example, to calculate the average triplet loss, we apply, consecutively, the model to A, then to P, then to N, and then we compute the loss for that example using eq. 7.3. We repeat that for all triplets in the batch and then compute the cost; gradient descent with backpropagation propagates the cost through the network to update its parameters.

It's a common misconception that for one-shot learning we need only one example of each entity for training. In practice, we need more than one example of each person for the person identification model to be accurate. It's called one-shot because of the most frequent application of such a model: face-based authentication. For example, such a model could be used to unlock your phone. If your model is good, then you only need to have *one picture* of you on your phone and it will recognize you, and also it will recognize that someone else is not you. When we have the model, to decide whether two pictures A and $\hat{A}$ belong to the same person, we check if $\|f(A) - f(\hat{A})\|^2$ is less than τ, a hyperparameter.

7.11 Zero-Shot Learning

I finish this chapter with **zero-shot learning**. It is a relatively new research area, so there are no algorithms that proved to have a significant practical utility yet. Therefore, I only outline here the basic idea and leave the details of various algorithms for further reading. In zero-shot learning (ZSL) we want to train a model to assign labels to objects. The most frequent application is to learn to assign labels to images.

However, contrary to standard classification, we want the model to be able to predict labels that we didn't have in the training data. How is that possible?

The trick is to use embeddings not just to represent the input **x** but also to represent the output y. Imagine that we have a model that for any word in English can generate an embedding vector with the following property: if a word y_i has a similar meaning to the word y_k, then the embedding vectors for these two words will be similar. For example, if y_i is *Paris* and y_k is *Rome*, then they will have embeddings that are similar; on the other hand, if y_k is *potato*, then the embeddings of y_i and y_k will be dissimilar. Such embedding vectors are called **word embeddings**, and they are usually compared using cosine similarity metrics[1].

---

[1]I will show in Chapter 10 how to learn words embeddings from data.

Word embeddings have such a property that each dimension of the embedding represents a specific feature of the meaning of the word. For example, if our word embedding has four dimensions (usually they are much wider, between 50 and 300 dimensions), then these four dimensions could represent such features of the meaning as *animalness*, *abstractness*, *sourness*, and *yellowness* (yes, sounds funny, but it's just an example). So the word *bee* would have an embedding like this $[1, 0, 0, 1]$, the word *yellow* like this $[0, 1, 0, 1]$, the word *unicorn* like this $[1, 1, 0, 0]$. The values for each embedding are obtained using a specific training procedure applied to a vast text corpus.

Now, in our classification problem, we can replace the label y_i for each example i in our training set with its word embedding and train a multi-label model that predicts word embeddings. To get the label for a new example $\mathbf{x}$, we apply our model f to $\mathbf{x}$, get the embedding $\hat{y}$ and then search among all English words those whose embeddings are the most similar to $\hat{y}$ using cosine similarity.

Why does that work? Take a zebra for example. It is white, it is a mammal, and it has stripes. Take a clownfish: it is orange, not a mammal, and has stripes. Now take a tiger: it is orange, it has stripes, and it is a mammal. If these three features are present in word embeddings, the CNN would learn to detect these same features in pictures. Even if the label *tiger* was absent in the training data, but other objects including zebras and clownfish were, then the CNN will most likely learn the notion of *mammalness*, *orangeness*, and *stripeness* to predict labels of those objects. Once we present the picture of a tiger to the model, those features will be correctly identified from the image and most likely the closest word embedding from our English dictionary to the predicted embedding will be that of *tiger*.

Chapter 8

Advanced Practice

This chapter contains the description of techniques that you could find useful in your practice in some contexts. It's called "Advanced Practice" not because the presented techniques are more complex, but rather because they are applied in some very specific contexts. In many practical situations, you will most likely not need to resort to using these techniques, but sometimes they are very helpful.

8.1 Handling Imbalanced Datasets

Often in practice, examples of some class will be underrepresented in your training data. This is the case, for example, when your classifier has to distinguish between genuine and fraudulent e-commerce transactions: the examples of genuine transactions are much more frequent. If you use SVM with soft margin, you can define a cost for misclassified examples. Because noise is always present in the training data, there are high chances that many examples of genuine transactions would end up on the wrong side of the decision boundary by contributing to the cost.

The SVM algorithm tries to move the hyperplane to avoid misclassified examples as much as possible. The "fraudulent" examples, which are in the minority, risk being misclassified in order to classify more numerous examples of the majority class correctly. This situation is illustrated in Figure 8.1a. This problem is observed for most learning algorithms applied to **imbalanced datasets**.

If you set the cost of misclassification of examples of the minority class higher, then the model will try harder to avoid misclassifying those examples, but this will incur the cost of misclassification of some examples of the majority class, as illustrated in Figure 8.1b.

Some SVM implementations allow you to provide weights for every class. The learning algorithm takes this information into account when looking for the best hyperplane.

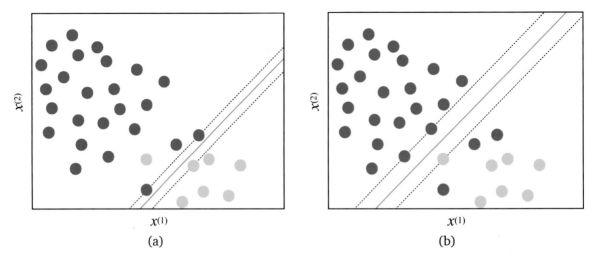

Figure 8.1: An illustration of an imbalanced problem. (a) Both classes have the same weight; (b) examples of the minority class have a higher weight.

If a learning algorithm doesn't allow weighting classes, you can try the technique of **oversampling**. It consists of increasing the importance of examples of some class by making multiple copies of the examples of that class.

An opposite approach, **undersampling**, is to randomly remove from the training set some examples of the majority class.

You might also try to create synthetic examples by randomly sampling feature values of several examples of the minority class and combining them to obtain a new example of that class. There are two popular algorithms that oversample the minority class by creating synthetic examples: the *synthetic minority oversampling technique* (**SMOTE**) and the *adaptive synthetic sampling method* (**ADASYN**).

SMOTE and ADASYN work similarly in many ways. For a given example $\mathbf{x}_i$ of the minority class, they pick k nearest neighbors of this example (let's denote this set of k examples $\mathcal{S}_k$) and then create a synthetic example $\mathbf{x}_{new}$ as $\mathbf{x}_i + \lambda(\mathbf{x}_{zi} - \mathbf{x}_i)$, where $\mathbf{x}_{zi}$ is an example of the minority class chosen randomly from $\mathcal{S}_k$. The interpolation hyperparameter λ is a random number in the range $[0, 1]$.

Both SMOTE and ADASYN randomly pick all possible $\mathbf{x}_i$ in the dataset. In ADASYN, the number of synthetic examples generated for each $\mathbf{x}_i$ is proportional to the number of examples in $\mathcal{S}_k$ which are not from the minority class. Therefore, more synthetic examples are generated in the area where the examples of the minority class are rare.

Some algorithms are less sensitive to the problem of an imbalanced dataset. Decision trees, as well as random forest and gradient boosting, often perform well on imbalanced datasets.

8.2 Combining Models

Ensemble algorithms, like Random Forest, typically combine models of the same nature. They boost performance by combining hundreds of weak models. In practice, we can sometimes get an additional performance gain by combining strong models made with different learning algorithms. In this case, we usually use only two or three models.

Three typical ways to combine models are 1) averaging, 2) majority vote and 3) stacking.

Averaging works for regression as well as those classification models that return classification scores. You simply apply all your models—let's call them **base models**—to the input $\mathbf{x}$ and then average the predictions. To see if the averaged model works better than each individual algorithm, you test it on the validation set using a metric of your choice.

Majority vote works for classification models. You apply all your base models to the input $\mathbf{x}$ and then return the majority class among all predictions. In the case of a tie, you either randomly pick one of the classes, or, you return an error message (if the fact of misclassifying would incur a significant cost).

Stacking consists of building a meta-model that takes the output of base models as input. Let's say you want to combine classifiers f_1 and f_2, both predicting the same set of classes. To create a training example $(\hat{\mathbf{x}}_i, \hat{y}_i)$ for the stacked model, set $\hat{\mathbf{x}}_i = [f_1(\mathbf{x}), f_2(\mathbf{x})]$ and $\hat{y}_i = y_i$.

If some of your base models return not just a class, but also a score for each class, you can use these values as features too.

To train the stacked model, it is recommended to use examples from the training set and tune the hyperparameters of the stacked model using cross-validation.

Obviously, you have to make sure that your stacked model performs better on the validation set than each of the base models you stacked.

The reason that combining multiple models can bring better performance is that when several uncorrelated strong models agree they are more likely to agree on the correct outcome. The keyword here is "uncorrelated." Ideally, base models should be obtained using different features or using algorithms of a different nature — for example, SVMs and Random Forest. Combining different versions of the decision tree learning algorithm, or several SVMs with different hyperparameters, may not result in a significant performance boost.

8.3 Training Neural Networks

In neural network training, one challenging aspect is how to convert your data into the input the network can work with. If your input is images, first of all, you have to resize all images so that they have the same dimensions. After that, pixels are usually first standardized and then normalized to the range $[0, 1]$.

Texts have to be tokenized (that is, split into pieces, such as words, punctuation marks, and other symbols). For CNN and RNN, each token is converted into a vector using the one-hot encoding, so the text becomes a list of one-hot vectors. Another, often better way to represent tokens is by using **word embeddings**. For a multilayer perceptron, to convert texts to vectors the bag of words approach may work well, especially for larger texts (larger than SMS messages and tweets).

The choice of specific neural network architecture is a difficult one. For the same problem, like seq2seq learning, there is a variety of architectures, and new ones are proposed almost every year. I recommend researching state of the art solutions for your problem using Google Scholar or Microsoft Academic search engines that allow searching for scientific publications using keywords and time range. If you don't mind working with less modern architecture, I recommend looking for implemented architectures on GitHub and finding one that could be applied to your data with minor modifications.

In practice, the advantage of a modern architecture over an older one becomes less significant as you preprocess, clean and normalize your data, and create a larger training set. Modern neural network architectures are a result of the collaboration of scientists from several labs and companies; such models could be very complex to implement on your own and usually require much computational power to train. Time spent trying to replicate results from a recent scientific paper may not be worth it. This time could better be spent on building the solution around a less modern but stable model and getting more training data.

Once you decided on the architecture of your network, you have to decide on the number of layers, their type, and size. It is recommended to start with one or two layers, train a model and see if it fits the training data well (has a low bias). If not, gradually increase the size of each layer and the number of layers until the model perfectly fits the training data. Once this is the case, if the model doesn't perform well on the validation data (has a high variance), you should add regularization to your model. If, after adding regularization, the model doesn't fit the training data anymore, slightly increase the size of the network. Continue iteratively until the model fits both training and validation data well enough according to your metric.

8.4 Advanced Regularization

In neural networks, besides L1 and L2 regularization, you can use neural network specific regularizers: **dropout, early stopping**, and **batch-normalization**. The latter is technically not a regularization technique, but it often has a regularization effect on the model.

The concept of dropout is very simple. Each time you run a training example through the network, you temporarily exclude at random some units from the computation. The higher the percentage of units excluded the higher the regularization effect. Neural network libraries allow you to add a dropout layer between two successive layers, or you can specify the dropout parameter for the layer. The dropout parameter is in the range $[0, 1]$ and it has to be found experimentally by tuning it on the validation data.

Early stopping is the way to train a neural network by saving the preliminary model after every epoch and assessing the performance of the preliminary model on the validation set. As you remember from the section about gradient descent in Chapter 4, as the number of epochs increases, the cost decreases. The decreased cost means that the model fits the training data well. However, at some point, after some epoch e, the model can start overfitting: the cost keeps decreasing, but the performance of the model on the validation data deteriorates. If you keep, in a file, the version of the model after each epoch, you can stop the training once you start observing a decreased performance on the validation set. Alternatively, you can keep running the training process for a fixed number of epochs and then, in the end, you pick the best model. Models saved after each epoch are called **checkpoints**. Some machine learning practitioners rely on this technique very often; others try to properly regularize the model to avoid such an undesirable behavior.

Batch normalization (which rather has to be called batch standardization) is a technique that consists of standardizing the outputs of each layer before the units of the subsequent layer receive them as input. In practice, batch normalization results in faster and more stable training, as well as some regularization effect. So it's always a good idea to try to use batch normalization. In neural network libraries, you can often insert a batch normalization layer between two layers.

Another regularization technique that can be applied not just to neural networks, but to virtually any learning algorithm, is called **data augmentation**. This technique is often used to regularize models that work with images. Once you have your original labeled training set, you can create a synthetic example from an original example by applying various transformations to the original image: zooming it slightly, rotating, flipping, darkening, and so on. You keep the original label in these synthetic examples. In practice, this often results in increased performance of the model.

8.5   Handling Multiple Inputs

Often in practice, you will work with multimodal data. For example, your input could be an image and text and the binary output could indicate whether the text describes this image.

It's hard to adapt **shallow learning** algorithms to work with multimodal data. However, it's not impossible. You could train one shallow model on the image and another one on the text. Then you can use a model combination technique we discussed above.

If you cannot divide your problem into two independent subproblems, you can try to vectorize each input (by applying the corresponding feature engineering method) and then simply concatenate two feature vectors together to form one wider feature vector. For example, if your image has features $[i^{(1)}, i^{(2)}, i^{(3)}]$ and your text has features $[t^{(1)}, t^{(2)}, t^{(3)}, t^{(4)}]$ your concatenated feature vector will be $[i^{(1)}, i^{(2)}, i^{(3)}, t^{(1)}, t^{(2)}, t^{(3)}, t^{(4)}]$.

With neural networks, you have more flexibility. You can build two subnetworks, one for each type of input. For example, a CNN subnetwork would read the image while an RNN subnetwork would read the text. Both subnetworks have as their last layer an embedding: CNN has an embedding of the image, while RNN has an embedding of the text. You can now concatenate two embeddings and then add a classification layer, such as softmax or sigmoid, on top of the concatenated embeddings. Neural network libraries provide simple-to-use tools that allow concatenating or averaging of layers from several subnetworks.

8.6 Handling Multiple Outputs

In some problems, you would like to predict multiple outputs for one input. We considered multi-label classification in the previous chapter. Some problems with multiple outputs can be effectively converted into a multi-label classification problem. Especially those that have labels of the same nature (like tags) or fake labels can be created as a full enumeration of combinations of original labels.

However, in some cases the outputs are multimodal, and their combinations cannot be effectively enumerated. Consider the following example: you want to build a model that detects an object on an image and returns its coordinates. In addition, the model has to return a tag describing the object, such as "person," "cat," or "hamster." Your training example will be a feature vector that represents an image. The label will be represented as a vector of coordinates of the object and another vector with a one-hot encoded tag.

To handle a situation like that, you can create one subnetwork that would work as an encoder. It will read the input image using, for example, one or several convolution layers. The encoder's last layer would be the embedding of the image. Then you add two other subnetworks on top of the embedding layer: one that takes the embedding vector as input and predicts the coordinates of an object. This first subnetwork can have a ReLU as the last layer, which is a good choice for predicting positive real numbers, such as coordinates; this subnetwork could use the mean squared error cost C_1. The second subnetwork will take the same embedding vector as input and predict the probabilities for each tag. This second subnetwork can have a softmax as the last layer, which is appropriate for the probabilistic output, and use the averaged negative log-likelihood cost C_2 (also called **cross-entropy** cost).

Obviously, you are interested in both accurately predicted coordinates and the tags. However, it is impossible to optimize the two cost functions at the same time. By trying to optimize one, you risk hurting the second one and the other way around. What you can do is add another hyperparameter γ in the range $(0, 1)$ and define the combined cost function as $\gamma C_1 + (1-\gamma)C_2$. Then you tune the value for γ on the validation data just like any other hyperparameter.

8.7 Transfer Learning

Transfer learning is probably where neural networks have a unique advantage over the shallow models. In transfer learning, you pick an existing model trained on some dataset, and you adapt this model to predict examples from another dataset, different from the one the model was built on. This second dataset is not like holdout sets you use for validation and test. It may represent some other phenomenon, or, as machine learning scientists say, it may come from another statistical distribution.

For example, imagine you have trained your model to recognize (and label) wild animals on a big labeled dataset. After some time, you have another problem to solve: you need to build a model that would recognize domestic animals. With shallow learning algorithms, you do not have many options: you have to build another big labeled dataset, now for domestic animals.

With neural networks, the situation is much more favorable. Transfer learning in neural networks works like this:

1. You build a deep model on the original big dataset (wild animals).
2. You compile a much smaller labeled dataset for your second model (domestic animals).
3. You remove the last one or several layers from the first model. Usually, these are layers responsible for the classification or regression; they usually follow the embedding layer.
4. You replace the removed layers with new layers adapted for your new problem.
5. You "freeze" the parameters of the layers remaining from the first model.
6. You use your smaller labeled dataset and gradient descent to train the parameters of only the new layers.

Usually, there is an abundance of deep models for visual problems available online. You can find one that has high chances to be of use for your problem, download that model, remove several last layers (the quantity of layers to remove is a hyperparameter), add your own prediction layers and train your model.

Even if you don't have an existing model, transfer learning can still help you in situations when your problem requires a labeled dataset that is very costly to obtain, but you can get another dataset for which labels are more readily available. Let's say you build a document classification model. You got the taxonomy of labels from your employer, and it contains a thousand categories. In this case, you would need to pay someone to a) read, understand and memorize the differences between categories and b) read up to a million documents and annotate them.

To save on labeling so many examples, you could consider using Wikipedia pages as the dataset to build your first model. The labels for a Wikipedia page can be obtained automatically by taking the category the Wikipedia page belongs to. Once your first model has learned to predict Wikipedia categories, you can "fine tune" this model to predict the categories of your employer's taxonomy. You will need much fewer annotated examples for your employer's problem than you would need if you started solving your original problem from scratch.

8.8 Algorithmic Efficiency

Not all algorithms capable of solving a problem are practical. Some can be too slow. Some problems can be solved by a fast algorithm; for others, no fast algorithms can exist.

The subfield of computer science called *analysis of algorithms* is concerned with determining and comparing the complexity of algorithms. **Big O notation** is used to classify algorithms according to how their running time or space requirements grow as the input size grows.

For example, let's say we have the problem of finding the two most distant one-dimensional examples in the set of examples S of size N. One algorithm we could craft to solve this problem would look like this (here and below, in Python):

```python
def find_max_distance(S):
    result = None
    max_distance = 0
    for x1 in S:
        for x2 in S:
            if abs(x1 - x2) >= max_distance:
                max_distance = abs(x1 - x2)
                result = (x1, x2)
    return result
```

In the above algorithm, we loop over all values in S, and at every iteration of the first loop, we loop over all values in S once again. Therefore, the above algorithm makes N^2 comparisons of numbers. If we take as a unit time the time the comparison, abs and assignment operations take, then the time complexity (or, simply, complexity) of this algorithm is at most $5N^2$. (At each iteration, we have one comparison, two abs and two assignment operations.) When the complexity of an algorithm is measured in the worst case, big O notation is used. For the above algorithm, using big O notation, we write that the algorithm's complexity is $O(N^2)$; the constants, like 5, are ignored.

For the same problem, we can craft another algorithm like this:

```python
def find_max_distance(S):
    result = None
    min_x = float("inf")
    max_x = float("-inf")
    for x in S:
        if x < min_x:
            min_x = x
        elif x > max_x:
            max_x = x
    result = (max_x, min_x)
    return result
```

In the above algorithm, we loop over all values in S only once, so the algorithm's complexity is $O(N)$. In this case, we say that the latter algorithm is *more efficient* than the former.

An algorithm is called efficient when its complexity is polynomial in the size of the input. Therefore both $O(N)$ and $O(N^2)$ are efficient because N is a polynomial of degree 1 and N^2 is a polynomial of degree 2. However, for very large inputs, an $O(N^2)$ algorithm can be slow. In the big data era, scientists often look for $O(\log N)$ algorithms.

From a practical standpoint, when you implement your algorithm, you should *avoid using loops whenever possible*. For example, you should use operations on matrices and vectors, instead of loops. In Python, to compute **wx**, you should write,

```
import numpy
wx = numpy.dot(w,x)
```

and not,

```
wx = 0
for i in range(N):
    wx += w[i]*x[i]
```

Use appropriate data structures. If the order of elements in a collection doesn't matter, use set instead of list. In Python, the operation of verifying whether a specific example x belongs to S is efficient when S is declared as a set and is inefficient when S is declared as a list.

Another important data structure that you can use to make your Python code more efficient is dict. It is called a dictionary or a hashmap in other languages. It allows you to define a collection of key-value pairs with very fast lookups for keys.

Unless you know exactly what you do, always prefer using popular libraries to writing your own scientific code. Scientific Python packages like numpy, scipy, and scikit-learn were built by experienced scientists and engineers with efficiency in mind. They have many methods implemented in the C programming language for maximum efficiency.

If you need to iterate over a vast collection of elements, use *generators* that create a function that returns one element at a time rather than all the elements at once.

Use the *cProfile* package in Python to find inefficiencies in your code.

Finally, when nothing can be improved in your code from the algorithmic perspective, you can further boost the speed of your code by using:

- *multiprocessing* package to run computations in parallel, and

- *PyPy*, *Numba* or similar tools to compile your Python code into fast, optimized machine code.

Chapter 9

Unsupervised Learning

Unsupervised learning deals with problems in which data doesn't have labels. That property makes it very problematic for many applications. The absence of labels representing the desired behavior for your model means the absence of a solid reference point to judge the quality of your model. In this book, I only present unsupervised learning methods that allow the building of models that can be evaluated based on data as opposed to human judgment.

9.1 Density Estimation

Density estimation is a problem of modeling the probability density function (pdf) of the unknown probability distribution from which the dataset has been drawn. It can be useful for many applications, in particular for novelty or intrusion detection. In Chapter 7, we already estimated the pdf to solve the one-class classification problem. To do that, we decided that our model would be **parametric**, more precisely a multivariate normal distribution (MVN). This decision was somewhat arbitrary because if the real distribution from which our dataset was drawn is different from the MVN, our model will be very likely far from perfect. We also know that models can be nonparametric. We used a **nonparametric model** in kernel regression. It turns out that the same approach can work for density estimation.

Let $\{x_i\}_{i=1}^N$ be a one-dimensional dataset (a multi-dimensional case is similar) whose examples were drawn from a distribution with an unknown pdf f with $x_i \in \mathbb{R}$ for all $i = 1, \ldots, N$. We are interested in modeling the shape of f. Our kernel model of f, denoted as $\hat{f}_b$, is given by,

$$\hat{f}_b(x) = \frac{1}{Nb} \sum_{i=1}^N k\left(\frac{x - x_i}{b}\right),\qquad(9.1)$$

where b is a hyperparameter that controls the tradeoff between bias and variance of our model and k is a kernel. Again, like in Chapter 7, we use a Gaussian kernel:

$$k(z) = \frac{1}{\sqrt{2\pi}} \exp\left(\frac{-z^2}{2}\right).$$

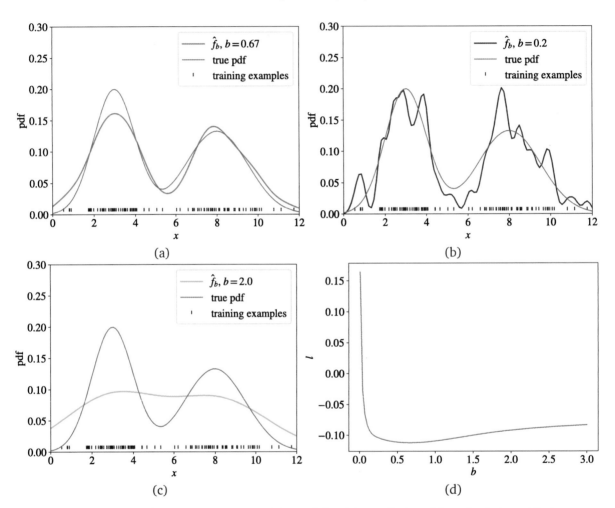

Figure 9.1: Kernel density estimation: (a) good fit; (b) overfitting; (c) underfitting; (d) the curve of grid search for the best value for b.

We look for such a value of b that minimizes the difference between the real shape of f and the shape of our model $\hat{f}_b$. A reasonable choice of measure of this difference is called the **mean integrated squared error** (MISE):

$$\text{MISE}(b) = \mathbb{E}\left[\int_{\mathbb{R}} (\hat{f}_b(x) - f(x))^2 \, dx\right]. \tag{9.2}$$

Intuitively, you see in eq. 9.2 that we square the difference between the real pdf f and our model of it $\hat{f}_b$. The integral $\int_{\mathbb{R}}$ replaces the summation $\sum_{i=1}^{N}$ we employed in the mean squared error, while the expectation operator $\mathbb{E}$ replaces the average $\frac{1}{N}$.

Indeed, when our loss is a function with a continuous domain, such as $(\hat{f}_b(x) - f(x))^2$, we have to replace the summation with the integral. The expectation operation $\mathbb{E}$ means that we want b to be optimal for all possible realizations of our training set $\{x_i\}_{i=1}^{N}$. That is important because $\hat{f}_b$ is defined on a *finite* sample of some probability distribution, while the real pdf f is defined on an infinite domain (the set $\mathbb{R}$).

Now, we can rewrite the right-hand side term in eq. 9.2 like this:

$$\mathbb{E}\left[\int_{\mathbb{R}} \hat{f}_b^2(x)dx\right] - 2\mathbb{E}\left[\int_{\mathbb{R}} \hat{f}_b(x)f(x)dx\right] + \mathbb{E}\left[\int_{\mathbb{R}} f(x)^2 dx\right].$$

The third term in the above summation is independent of b and thus can be ignored. An unbiased estimator of the first term is given by $\int_{\mathbb{R}} \hat{f}_b^2(x)dx$ while the unbiased estimator of the second term can be approximated by **cross-validation** $-\frac{2}{N} \sum_{i=1}^{N} \hat{f}_b^{(i)}(x_i)$, where $\hat{f}_b^{(i)}$ is a kernel model of f computed on our training set with the example x_i excluded.

The term $\sum_{i=1}^{N} \hat{f}_b^{(i)}(x_i)$ is known in statistics as the **leave one out estimate**, a form of cross-validation in which each fold consists of one example. You could have noticed that the term $\int_{\mathbb{R}} \hat{f}_b(x)f(x)dx$ (let's call it a) is the expected value of the function $\hat{f}_b$, because f is a pdf. It can be demonstrated that the leave one out estimate is an unbiased estimator of $\mathbb{E}[a]$.

Now, to find the optimal value b^* for b, we minimize the cost defined as,

$$\int_{\mathbb{R}} \hat{f}_b^2(x)dx - \frac{2}{N} \sum_{i=1}^{N} \hat{f}_b^{(i)}(x_i).$$

We can find b^* using grid search. For D-dimensional feature vectors $\mathbf{x}$, the error term $x - x_i$ in eq. 9.1 can be replaced by the Euclidean distance $\|\mathbf{x} - \mathbf{x}_i\|$. In Figure 9.1 you can see the estimates for the same pdf obtained with three different values of b from a 100-example dataset, as well as the grid search curve. We pick b^* at the minimum of the grid search curve.

9.2 Clustering

Clustering is a problem of learning to assign a label to examples by leveraging an unlabeled dataset. Because the dataset is completely unlabeled, deciding on whether the learned model is optimal is much more complicated than in supervised learning.

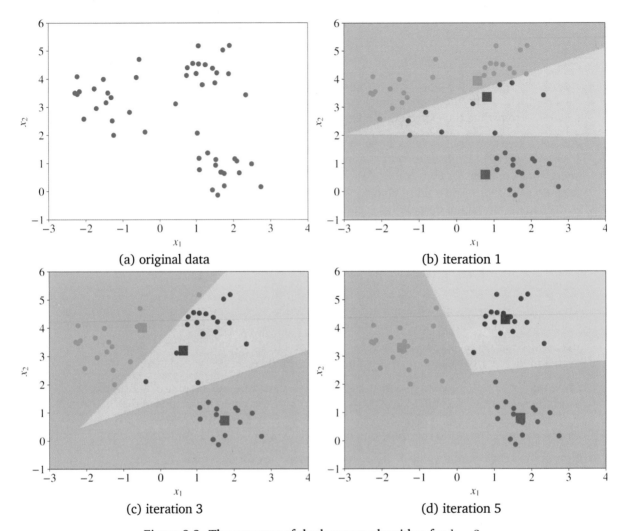

Figure 9.2: The progress of the k-means algorithm for $k = 3$.

There is a variety of clustering algorithms, and, unfortunately, it's hard to tell which one is better in quality for your dataset. Usually, the performance of each algorithm depends on the unknown properties of the probability distribution that the dataset was drawn from. In this Chapter, I outline the most useful and widely used clustering algorithms.

9.2.1 K-Means

The **k-means** clustering algorithm works as follows. First, you choose k — the number of clusters. Then you randomly put k feature vectors, called **centroids**, to the feature space.

We then compute the distance from each example **x** to each centroid **c** using some metric, like the Euclidean distance. Then we assign the closest centroid to each example (like if we labeled each example with a centroid id as the label). For each centroid, we calculate the average feature vector of the examples labeled with it. These average feature vectors become the new locations of the centroids.

We recompute the distance from each example to each centroid, modify the assignment and repeat the procedure until the assignments don't change after the centroid locations were recomputed. The model is the list of assignments of centroids IDs to the examples.

The initial position of centroids influence the final positions, so two runs of k-means can result in two different models. Some variants of k-means compute the initial positions of centroids based on some properties of the dataset.

One run of the k-means algorithm is illustrated in Figure 9.2. The circles in Figure 9.2 are two-dimensional feature vectors; the squares are moving centroids. Different background colors represent regions in which all points belong to the same cluster.

The value of k, the number of clusters, is a hyperparameter that has to be tuned by the data analyst. There are some techniques for selecting k. None of them is proven optimal. Most of those techniques require the analyst to make an "educated guess" by looking at some metrics or by examining cluster assignments visually. In this chapter, I present one approach to choose a reasonably good value for k without looking at the data and making guesses.

9.2.2 DBSCAN and HDBSCAN

While k-means and similar algorithms are centroid-based, **DBSCAN** is a density-based clustering algorithm. Instead of guessing how many clusters you need, by using DBSCAN, you define two hyperparameters: ϵ and n. You start by picking an example **x** from your dataset at random and assign it to cluster 1. Then you count how many examples have the distance from **x** less than or equal to ϵ. If this quantity is greater than or equal to n, then you put all these ϵ-neighbors to the same cluster 1. You then examine each member of cluster 1 and find their respective ϵ-neighbors. If some member of cluster 1 has n or more ϵ-neighbors, you expand cluster 1 by adding those ϵ-neighbors to the cluster. You continue expanding cluster 1 until there are no more examples to put in it. In the latter case, you pick from the dataset another example not belonging to any cluster and put it to cluster 2. You continue like this until all examples either belong to some cluster or are marked as outliers. An outlier is an example whose ϵ-neighborhood contains less than n examples.

The advantage of DBSCAN is that it can build clusters that have an arbitrary shape, while k-means and other centroid-based algorithms create clusters that have a shape of a hypersphere. An obvious drawback of DBSCAN is that it has two hyperparameters and choosing good values for them (especially ϵ) could be challenging. Furthermore, having ϵ fixed, the clustering algorithm cannot effectively deal with clusters of varying density.

HDBSCAN is the clustering algorithm that keeps the advantages of DBSCAN, by removing the need to decide on the value of ϵ. The algorithm is capable of building clusters of varying density. HDBSCAN is an ingenious combination of multiple ideas and describing the algorithm in full is beyond the scope of this book.

HDBSCAN only has one important hyperparameter: n, the minimum number of examples to put in a cluster. This hyperparameter is relatively simple to choose by intuition. HDBSCAN has very fast implementations: it can deal with millions of examples effectively. Modern implementations of k-means are much faster than HDBSCAN, though, but the qualities of the latter may outweigh its drawbacks for many practical tasks. I recommend to always trying HDBSCAN on your data first.

9.2.3 Determining the Number of Clusters

The most important question is how many clusters does your dataset have? When the feature vectors are one-, two- or three-dimensional, you can look at the data and see "clouds" of points in the feature space. Each cloud is a potential cluster. However, for D-dimensional data, with $D > 3$, looking at the data is problematic[1].

One way of determining the reasonable number of clusters is based on the concept of **prediction strength**. The idea is to split the data into training and test set, similarly to how we do in supervised learning. Once you have the training and test sets, $\mathcal{S}_{tr}$ of size N_{tr} and $\mathcal{S}_{te}$ of size N_{te} respectively, you fix k, the number of clusters, and run a clustering algorithm C on sets $\mathcal{S}_{tr}$ and $\mathcal{S}_{te}$ and obtain the clustering results $C(\mathcal{S}_{tr}, k)$ and $C(\mathcal{S}_{te}, k)$.

Let A be the clustering $C(\mathcal{S}_{tr}, k)$ built using the training set. The clusters in A can be seen as regions. If an example falls within one of those regions, then that example belongs to some specific cluster. For example, if we apply the k-means algorithm to some dataset, it results in a partition of the feature space into k polygonal regions, as we saw in Figure 9.2.

Define the $N_{te} \times N_{te}$ **co-membership matrix** $\mathbf{D}[A, \mathcal{S}_{te}]$ as follows: $\mathbf{D}[A, \mathcal{S}_{te}]^{(i,i')} = 1$ if and only if examples $\mathbf{x}_i$ and $\mathbf{x}_{i'}$ from the test set belong to the same cluster according to the clustering A. Otherwise $\mathbf{D}[A, \mathcal{S}_{te}]^{(i,i')} = 0$.

Let's take a break and see what we have here. We have built, *using the training set* of examples, a clustering A that has k clusters. Then we have built the co-membership matrix that indicates whether two examples *from the test set* belong to the same cluster in A.

Intuitively, if the quantity k is the reasonable number of clusters, then two examples that belong to the same cluster in clustering $C(\mathcal{S}_{te}, k)$ will most likely belong to the same cluster in clustering $C(\mathcal{S}_{tr}, k)$. On the other hand, if k is not reasonable (too high or too low), then training data-based and test data-based clusterings will likely be less consistent.

---

[1]Some analysts look at multiple two-dimensional plots, in which only a pair of features are present at the same time. It might give an intuition about the number of clusters. However, such an approach suffers from subjectivity, is prone to error and counts as an educated guess rather than a scientific method.

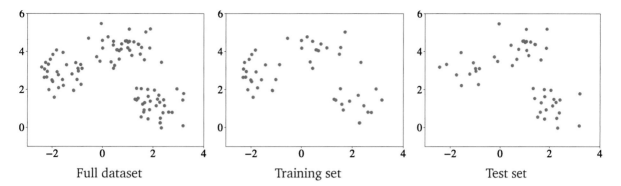

Figure 9.3: Data used for clustering illustrated in Figure 9.4

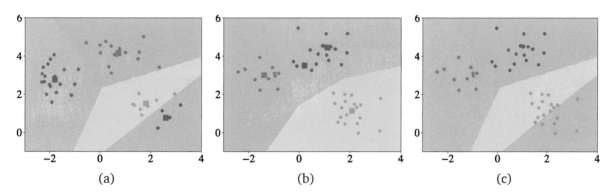

Figure 9.4: The clustering for $k = 4$: (a) training data clustering; (b) test data clustering; (c) test data plotted over the training clustering.

Using the data shown in Figure 9.3, the idea is illustrated in Figure 9.4. The plots in Figure 9.4a and 9.4b show respectively $C(\mathcal{S}_{tr}, 4)$ and $C(\mathcal{S}_{te}, 4)$ with their respective cluster regions. Figure 9.4c shows the test examples plotted over the training data cluster regions. You can see in 9.4c that orange test examples don't belong anymore to the same cluster according to the clustering regions obtained from the training data. This will result in many zeroes in the matrix $\mathbf{D}[A, \mathcal{S}_{te}]$ which, in turn, is an indicator that $k = 4$ is likely not the best number of clusters.

More formally, the prediction strength for the number of clusters k is given by,

$$\mathrm{ps}(k) \overset{\text{def}}{=} \min_{j=1,\ldots,k} \frac{1}{|A_j|(|A_j| - 1)} \sum_{i,i' \in A_j} \mathbf{D}[A, \mathcal{S}_{te}]^{(i,i')},$$

where $A \overset{\text{def}}{=} C(\mathcal{S}_{tr}, k)$, A_j is j^{th} cluster from the clustering $C(\mathcal{S}_{te}, k)$ and $|A_j|$ is the number of examples in cluster A_j.

Given a clustering $C(\mathcal{S}_{tr}, k)$, for each test cluster, we compute the proportion of observation pairs in that cluster that are also assigned to the same cluster by the training set centroids. The prediction strength is the minimum of this quantity over the k test clusters.

Experiments suggest that a reasonable number of clusters is the largest k such that $\mathrm{ps}(k)$ is above 0.8. You can see in Figure 9.5 examples of predictive strength for different values of k for two-, three- and four-cluster data.

For non-deterministic clustering algorithms, such as k-means, which can generate different clusterings depending on the initial positions of centroids, it is recommended to do multiple runs of the clustering algorithm for the same k and compute the average prediction strength $\bar{\mathrm{ps}}(k)$ over multiple runs.

Another effective method to estimate the number of clusters is the **gap statistic** method. Other, less automatic methods, which some analysts still use, include the **elbow method** and the **average silhouette method**.

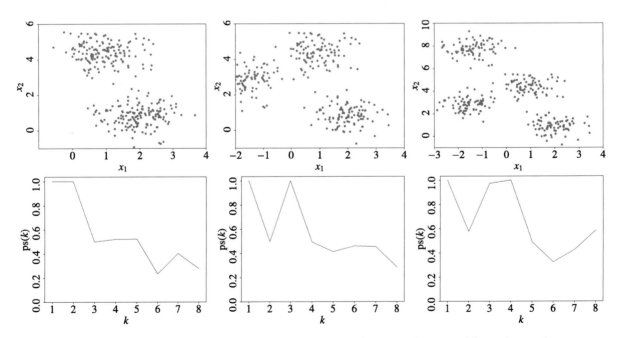

Figure 9.5: Predictive strength for different values of k for two-, three- and four-cluster data.

9.2.4 Other Clustering Algorithms

DBSCAN and k-means compute so-called **hard clustering**, in which each example can belong to only one cluster. **Gaussian mixture model** (GMM) allows each example to be a member of several clusters with different *membership score* (HDBSCAN also allows this). Computing a GMM is very similar to doing model-based density estimation. In GMM, instead of having just one multivariate normal distribution (MND), we have a weighted sum of several MNDs:

$$f_X = \sum_{j=1}^{k} \phi_j f_{\boldsymbol{\mu}_j, \boldsymbol{\Sigma}_j},$$

where $f_{\boldsymbol{\mu}_j, \boldsymbol{\Sigma}_j}$ is a MND j, and ϕ_j is its weight in the sum. The values of parameters $\boldsymbol{\mu}_j$, $\boldsymbol{\Sigma}_j$, and ϕ_j, for all $j = 1, \ldots, k$ are obtained using the **expectation maximization algorithm** (EM) to optimize the **maximum likelihood** criterion.

Again, for simplicity, let us look at the one-dimensional data. Also assume that there are two clusters: $k = 2$. In this case, we have two Gaussian distributions,

$$f(x \mid \mu_1, \sigma_1^2) = \frac{1}{\sqrt{2\pi\sigma_1^2}} \exp{-\frac{(x - \mu_1)^2}{2\sigma_1^2}} \text{ and } f(x \mid \mu_2, \sigma_2^2) = \frac{1}{\sqrt{2\pi\sigma_2^2}} \exp{-\frac{(x - \mu_2)^2}{2\sigma_2^2}}, \quad (9.3)$$

where $f(x \mid \mu_1, \sigma_1^2)$ and $f(x \mid \mu_2, \sigma_2^2)$ are two pdf defining the likelihood of $X = x$.

We use the EM algorithm to estimate μ_1, σ_1^2, μ_2, σ_2^2, ϕ_1, and ϕ_2. The parameters ϕ_1 and ϕ_2 are useful for the density estimation and less useful for clustering, as we will see below.

EM works as follows. In the beginning, we guess the initial values for μ_1, σ_1^2, μ_2, and σ_2^2, and set $\phi_1 = \phi_2 = \frac{1}{2}$ (in general, it's $\frac{1}{k}$ for each ϕ_j, $j \in 1, \ldots, k$).

At each iteration of EM, the following four steps are executed:

1. For all $i = 1, \ldots, N$, calculate the likelihood of each x_i using eq. 9.3:

$$f(x_i \mid \mu_1, \sigma_1^2) \leftarrow \frac{1}{\sqrt{2\pi\sigma_1^2}} \exp{-\frac{(x_i - \mu_1)^2}{2\sigma_1^2}} \text{ and } f(x_i \mid \mu_2, \sigma_2^2) \leftarrow \frac{1}{\sqrt{2\pi\sigma_2^2}} \exp{-\frac{(x_i - \mu_2)^2}{2\sigma_2^2}}.$$

2. Using **Bayes' Rule**, for each example x_i, calculate the likelihood $b_i^{(j)}$ that the example belongs to cluster $j \in \{1, 2\}$ (in other words, the likelihood that the example was drawn from the Gaussian j):

$$b_i^{(j)} \leftarrow \frac{f(x_i \mid \mu_j, \sigma_j^2)\phi_j}{f(x_i \mid \mu_1, \sigma_1^2)\phi_1 + f(x_i \mid \mu_2, \sigma_2^2)\phi_2}.$$

The parameter ϕ_j reflects how likely is that our Gaussian distribution j with parameters μ_j and σ_j^2 may have produced our dataset. That is why in the beginning we set $\phi_1 = \phi_2 = \frac{1}{2}$: we don't know how each of the two Gaussians is likely, and we reflect our ignorance by setting the likelihood of both to one half.

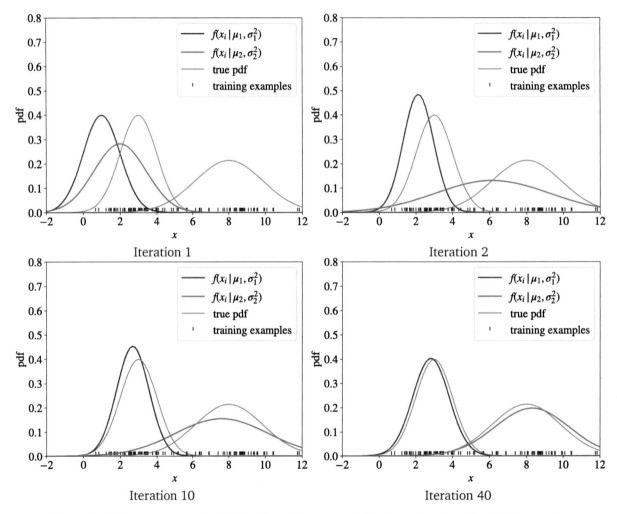

Figure 9.6: The progress of the Gaussian mixture model estimation using the EM algorithm for two clusters ($k = 2$).

3. Compute the new values of μ_j and σ_j^2, $j \in \{1, 2\}$ as,

$$\mu_j \leftarrow \frac{\sum_{i=1}^{N} b_i^{(j)} x_i}{\sum_{i=1}^{N} b_i^{(j)}} \text{ and } \sigma_j^2 \leftarrow \frac{\sum_{i=1}^{N} b_i^{(j)} (x_i - \mu_j)^2}{\sum_{i=1}^{N} b_i^{(j)}}. \tag{9.4}$$

4. Update ϕ_j, $j \in \{1, 2\}$ as,

$$\phi_j \leftarrow \frac{1}{N} \sum_{i=1}^{N} b_i^{(j)}.$$

The steps $1 - 4$ are executed iteratively until the values μ_j and σ_j^2 don't change much: for example, the change is below some threshold ϵ. Figure 9.6 illustrates this process.

You may have noticed that the EM algorithm is very similar to the k-means algorithm: start with random clusters, then iteratively update each cluster's parameters by averaging the data that is assigned to that cluster. The only difference in the case of GMM is that the assignment of an example x_i to the cluster j is **soft**: x_i belongs to cluster j with probability $b_i^{(j)}$. This is why we calculate the new values for μ_j and σ_j^2 in eq. 9.4 not as an average (used in k-means) but as a **weighted average** with weights $b_i^{(j)}$.

Once we have learned the parameters μ_j and σ_j^2 for each cluster j, the membership score of example x in cluster j is given by $f(x \mid \mu_j, \sigma_j^2)$.

The extension to D-dimensional data ($D > 1$) is straightforward. The only difference is that instead of the variance σ^2, we now have the covariance matrix Σ that parametrizes the multinomial normal distribution (MND).

Contrary to k-means where clusters can only be circular, the clusters in GMM have the form of an ellipse that can have an arbitrary elongation and rotation. The values in the covariance matrix control these properties.

There's no universally recognized method to choose the right k in GMM. I recommend that you first split the dataset into training and test set. Then you try different k and build a different model f_{tr}^k for each k on the training data. You pick the value of k that maximizes the likelihood of examples in the test set:

$$\underset{k}{\arg \max} \prod_{i=1}^{|N_{te}|} f_{tr}^k(\mathbf{x}_i),$$

where $|N_{te}|$ is the size of the test set.

There is a variety of clustering algorithms described in the literature. Worth mentioning are **spectral clustering** and **hierarchical clustering**. For some datasets, you may find those more appropriate. However, in most practical cases, k-means, HDBSCAN and the Gaussian mixture model would satisfy your needs.

9.3 Dimensionality Reduction

Modern machine learning algorithms, such as ensemble algorithms and neural networks, handle well very high-dimensional examples, up to millions of features. With modern computers and graphical processing units (GPUs), **dimensionality reduction** techniques are used less in practice than in the past. The most frequent use case for dimensionality reduction is data visualization: humans can only interpret a maximum of three dimensions on a plot.

Another situation in which you could benefit from dimensionality reduction is when you have to build an interpretable model and to do so you are limited in your choice of learning algorithms. For example, you can only use decision tree learning or linear regression. By reducing your data to lower dimensionality and by figuring out which quality of the original example each new feature in the reduced feature space reflects, your can use simpler algorithms. Dimensionality reduction removes redundant or highly correlated features; it also reduces the noise in the data — all that contributes to the interpretability of the model.

Three widely used techniques of dimensionality reduction are **principal component analysis** (PCA), **uniform manifold approximation and projection** (UMAP), and **autoencoders**.

I already explained autoencoders in Chapter 7. You can use the low-dimensional output of the **bottleneck layer** of the autoencoder as the vector of reduced dimensionality that represents the high-dimensional input feature vector. You know that this low-dimensional vector represents the essential information contained in the input vector because the autoencoder is capable of reconstructing the input feature vector based on the bottleneck layer output alone.

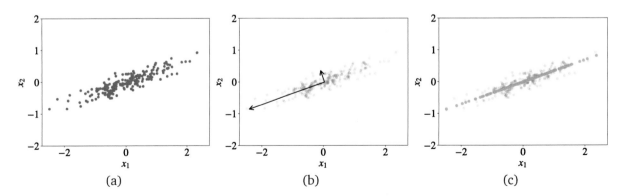

Figure 9.7: PCA: (a) the original data; (b) two principal components displayed as vectors; (c) the data projected on the first principal component.

9.3.1 Principal Component Analysis

Principal component analysis or PCA is one of the oldest dimensionality reduction methods. The math behind it involves operation on matrices that I didn't explain in Chapter 2, so I leave the math of PCA for your further reading. Here, I only provide intuition and illustrate the method on an example.

Consider a two-dimensional dataset as shown in Figure 9.7a. Principal components are vectors that define a new coordinate system in which the first axis goes in the direction of the highest variance in the data. The second axis is orthogonal to the first one and goes in the direction of the second highest variance in the data. If our data was three-dimensional, the third axis would be orthogonal to both the first and the second axes and go in the direction of the third highest variance, and so on. In Figure 9.7b, the two principal components are shown as arrows. The length of the arrow reflects the variance in this direction.

Now, if we want to reduce the dimensionality of our data to $D_{new} < D$, we pick D_{new} largest principal components and project our data points on them. For our two-dimensional illustration, we can set $D_{new} = 1$ and project our examples to the first principal component to obtain the orange points in Figure 9.7c.

To describe each orange point, we need only one coordinate instead of two: the coordinate with respect to the first principal component. When our data is very high-dimensional, it often happens in practice that the first two or three principal components account for most of the variation in the data, so by displaying the data on a 2D or 3D plot we can indeed see a very high-dimensional data and its properties.

9.3.2 UMAP

The idea behind many of the modern dimensionality reduction algorithms, especially those designed specifically for visualization purposes such as **t-SNE** and **UMAP**, is basically the same. We first design a similarity metric for two examples. For visualization purposes, besides the Euclidean distance between the two examples, this similarity metric often reflects some local properties of the two examples, such as the density of other examples around them.

In UMAP, this similarity metric w is defined as follows,

$$w(\mathbf{x}_i, \mathbf{x}_j) \stackrel{\text{def}}{=} w_i(\mathbf{x}_i, \mathbf{x}_j) + w_j(\mathbf{x}_j, \mathbf{x}_i) - w_i(\mathbf{x}_i, \mathbf{x}_j)w_j(\mathbf{x}_j, \mathbf{x}_i). \tag{9.5}$$

The function $w_i(\mathbf{x}_i, \mathbf{x}_j)$ is defined as,

$$w_i(\mathbf{x}_i, \mathbf{x}_j) \stackrel{\text{def}}{=} \exp\left(-\frac{d(\mathbf{x}_i, \mathbf{x}_j) - \rho_i}{\sigma_i}\right),$$

where $d(\mathbf{x}_i, \mathbf{x}_j)$ is the Euclidean distance between two examples, ρ_i is the distance from $\mathbf{x}_i$ to its closest neighbor, and σ_i is the distance from $\mathbf{x}_i$ to its kth closest neighbor (k is a hyperparameter of the algorithm).

It can be shown that the metric in eq. 9.5 varies in the range from 0 to 1 and is symmetric, which means that $w(\mathbf{x}_i, \mathbf{x}_j) = w(\mathbf{x}_j, \mathbf{x}_i)$.

Let w denote the similarity of two examples in the original high-dimensional space and let w' be the similarity given by the same eq. 9.5 in the new low-dimensional space.

To continue, I need to quickly introduce the notion of a **fuzzy set**. A fuzzy set is a generalization of a set. For each element x in a fuzzy set S, there's a membership function $\mu_S(x) \in [0, 1]$ that defines the *membership strength* of x to the set S. We say that x weakly belongs to a fuzzy set S if $\mu_S(x)$ is close to zero. On the other hand, if $\mu_S(x)$ is close to 1, then x has a strong membership in S. If $\mu(x) = 1$ for all $x \in S$, then a fuzzy set S becomes equivalent to a normal, nonfuzzy set.

Let's now see why we need this notion of a fuzzy set here.

Because the values of w and w' lie in the range between 0 and 1, we can see $w(\mathbf{x}_i, \mathbf{x}_j)$ as membership of the pair of examples $(\mathbf{x}_i, \mathbf{x}_j)$ in a certain fuzzy set. The same can be said about w'. The notion of similarity of two fuzzy sets is called **fuzzy set cross-entropy** and is defined as,

$$C_{w,w'} = \sum_{i=1}^{N} \sum_{j=1}^{N} \left[w(\mathbf{x}_i, \mathbf{x}_j) \ln \left(\frac{w(\mathbf{x}_i, \mathbf{x}_j)}{w'(\mathbf{x}_i', \mathbf{x}_j')} \right) + (1 - w(\mathbf{x}_i, \mathbf{x}_j)) \ln \left(\frac{1 - w(\mathbf{x}_i, \mathbf{x}_j)}{1 - w'(\mathbf{x}_i', \mathbf{x}_j')} \right) \right], \quad (9.6)$$

where $\mathbf{x}'$ is the low-dimensional "version" of the original high-dimensional example $\mathbf{x}$.

In eq. 9.6 the unknown parameters are $\mathbf{x}_i'$ (for all $i = 1, \ldots, N$), the low-dimensional examples we look for. We can compute them by gradient descent by minimizing $C_{w,w'}$.

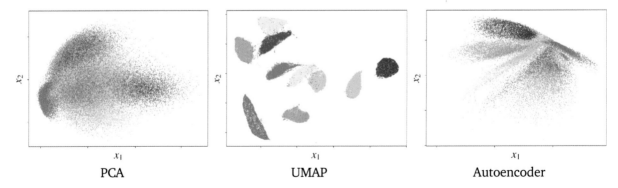

PCA UMAP Autoencoder

Figure 9.8: Dimensionality reduction of the MNIST dataset using three different techniques.

In Figure 9.8, you can see the result of dimensionality reduction applied to the MNIST dataset of handwritten digits. MNIST is commonly used for benchmarking various image processing systems; it contains 70,000 labeled examples. Ten different colors on the plot correspond to ten classes. Each point on the plot corresponds a specific example in the dataset. As you can see, UMAP separates examples visually better (remember, it doesn't have access to labels). In practice, UMAP is slightly slower than PCA but faster than autoencoder.

9.4 Outlier Detection

Outlier detection is the problem of detecting the examples in the dataset that are very different from what a typical example in the dataset looks like. We have already seen several techniques that could help to solve this problem: autoencoder and one-class classifier learning. If we use an autoencoder, we train it on our dataset. Then, if we want to predict whether an example is an outlier, we can use the autoencoder model to reconstruct the example from the bottleneck layer. The model will unlikely be capable of reconstructing an outlier.

In one-class classification, the model either predicts that the input example belongs to the class, or it's an outlier.

Chapter 10

Other Forms of Learning

10.1 Metric Learning

I mentioned that the most frequently used metrics of similarity (or dissimilarity) between two feature vectors are **Euclidean distance** and **cosine similarity**. Such choices of metric seem logical but arbitrary, just like the choice of the squared error in linear regression (or the form of linear regression itself). The fact that one metric can work better than another depending on the dataset is an indicator that none of them are perfect.

You can *create* a metric that would work better for your dataset. It's then possible to integrate your metric into any learning algorithm that needs a metric, like k-means or kNN. How can you know, without trying all possibilities, which equation would be a good metric? As you could already guess, a metric can be learned from data.

Remember the Euclidean distance between two feature vectors $\mathbf{x}$ and $\mathbf{x}'$:

$$d(\mathbf{x}, \mathbf{x}') = \|\mathbf{x} - \mathbf{x}'\| \overset{\text{def}}{=} \sqrt{(\mathbf{x} - \mathbf{x}')^2} = \sqrt{(\mathbf{x} - \mathbf{x}')(\mathbf{x} - \mathbf{x}')}.$$

We can slightly modify this metric to make it parametrizable and then learn these parameters from data. Consider the following modification:

$$d_{\mathbf{A}}(\mathbf{x}, \mathbf{x}') = \|\mathbf{x} - \mathbf{x}'\|_{\mathbf{A}} \overset{\text{def}}{=} \sqrt{(\mathbf{x} - \mathbf{x}')^{\top} \mathbf{A} (\mathbf{x} - \mathbf{x}')},$$

where $\mathbf{A}$ is a $D \times D$ matrix. Let's say $D = 3$. If we let $\mathbf{A}$ be the identity matrix,

$$\mathbf{A} \overset{\text{def}}{=} \begin{bmatrix} 1 & 0 & 0 \\ 0 & 1 & 0 \\ 0 & 0 & 1 \end{bmatrix},$$

then $d_{\mathbf{A}}$ becomes the Euclidean distance. If we have a general diagonal matrix, like this:

$$\mathbf{A} \stackrel{\text{def}}{=} \begin{bmatrix} 2 & 0 & 0 \\ 0 & 8 & 0 \\ 0 & 0 & 1 \end{bmatrix},$$

then different dimensions have different importance in the metric. (In the above example, the second dimension is the most important in the metric calculation.) More generally, to be called a metric a function of two variables has to satisfy three conditions:

1. $d(\mathbf{x}, \mathbf{x}') \geq 0$ nonnegativity,
2. $d(\mathbf{x}, \mathbf{x}') \leq d(\mathbf{x}, \mathbf{z}) + d(\mathbf{z}, \mathbf{x}')$ triangle inequality,
3. $d(\mathbf{x}, \mathbf{x}') = d(\mathbf{x}', \mathbf{x})$ symmetry.

To satisfy the first two conditions, the matrix $\mathbf{A}$ has to be *positive semidefinite*. You can see a positive semidefinite matrix as the generalization of the notion of a nonnegative real number to matrices. Any positive semidefinite matrix $\mathbf{M}$ satisfies:

$$\mathbf{z}^{\top} \mathbf{M} \mathbf{z} \geq 0,$$

for any vector $\mathbf{z}$ having the same dimensionality as the number of rows and columns in $\mathbf{M}$.

The above property follows from the definition of a positive semidefinite matrix. The proof that the second condition is satisfied when the matrix $\mathbf{A}$ is positive semidefinite can be found on the book's companion website.

To satisfy the third condition, we can simply take $(d(\mathbf{x}, \mathbf{x}') + d(\mathbf{x}', \mathbf{x}))/2$.

Let's say we have an unannotated set $\mathcal{X} = \{\mathbf{x}_i\}_{i=1}^{N}$. To build the training data for our metric learning problem, we manually create two sets. The first set $\mathcal{S}$ is such that a pair of examples $(\mathbf{x}_i, \mathbf{x}_k)$ belongs to set $\mathcal{S}$ if $\mathbf{x}_i$ and $\mathbf{x}_k$ are similar (from our subjective perspective). The second set $\mathcal{D}$ is such that a pair of examples $(\mathbf{x}_i, \mathbf{x}_k)$ belongs to set $\mathcal{D}$ if $\mathbf{x}_i$ and $\mathbf{x}_k$ are dissimilar.

To train the matrix of parameters $\mathbf{A}$ from the data, we want to find a positive semidefinite matrix $\mathbf{A}$ that solves the following optimization problem:

$$\min_{\mathbf{A}} \sum_{(\mathbf{x}_i, \mathbf{x}_k) \in \mathcal{S}} \|\mathbf{x} - \mathbf{x}'\|_{\mathbf{A}}^{2} \text{ such that } \sum_{(\mathbf{x}_i, \mathbf{x}_k) \in \mathcal{D}} \|\mathbf{x} - \mathbf{x}'\|_{\mathbf{A}} \geq c,$$

where c is a positive constant (can be any number).

The solution to this optimization problem is found by gradient descent with a modification that ensures that the found matrix $\mathbf{A}$ is positive semidefinite. We leave the description of the algorithm out of the scope of this book for further reading.

I should point out that **one-shot learning** with **siamese networks** and **triplet loss** can be seen as metric learning problem: the pairs of pictures of the same person belong to the set $\mathcal{S}$, while pairs of random pictures belong to $\mathcal{D}$.

There are many other ways to learn a metric, including non-linear and kernel-based. However, the one presented in this book, as well as the adaptation of one-shot learning, should suffice for most practical applications.

10.2   Learning to Rank

Learning to rank is a supervised learning problem. Among others, one frequent problem solved using learning to rank is the optimization of search results returned by a search engine for a query. In search result ranking optimization, a labeled example $\mathcal{X}_i$ in the training set of size N is a ranked collection of documents of size r_i (labels are ranks of documents). A feature vector represents each document in the collection. The goal of the learning is to find a ranking function f which outputs values that can be used to rank documents. For each training example, an ideal function f would output values that induce the same ranking of documents as given by the labels.

Each example $\mathcal{X}_i$, $i = 1, \ldots, N$, is a collection of feature vectors with labels: $\mathcal{X}_i = \{(\mathbf{x}_{i,j}, y_{i,j})\}_{j=1}^{r_i}$. Features in a feature vector $\mathbf{x}_{i,j}$ represent the document $j = 1, \ldots, r_i$. For example, $x_{i,j}^{(1)}$ could represent how recent is the document, $x_{i,j}^{(2)}$ would reflect whether the words of the query can be found in the document title, $x_{i,j}^{(3)}$ could represent the size of the document, and so on. The label $y_{i,j}$ could be the rank $(1, 2, \ldots, r_i)$ or a score. For example, the lower the score, the higher the document should be ranked.

There are three approaches to solve that problem: **pointwise**, **pairwise**, and **listwise**.

The pointwise approach transforms each training example into multiple examples: one example per document. The learning problem becomes a standard supervised learning problem, either regression or logistic regression. In each example $(\mathbf{x}, y)$ of the pointwise learning problem, $\mathbf{x}$ is the feature vector of some document, and y is the original score (if $y_{i,j}$ is a score) or a synthetic score obtained from the ranking (the higher the rank, the lower the synthetic score). Any supervised learning algorithm can be used in this case. The solution is usually far from perfect. Principally, this is because each document is considered in isolation, while the original ranking (given by the labels $y_{i,j}$ of the original training set) could optimize the positions of the whole set of documents. For example, if we have already given a high rank to a Wikipedia page in some collection of documents, we would prefer not giving a high rank to another Wikipedia page for the same query.

In the pairwise approach, we also consider documents in isolation, but, in this case, a pair of documents is considered at once. Given a pair of documents $(\mathbf{x}_i, \mathbf{x}_k)$ we build a model f, which, given $(\mathbf{x}_i, \mathbf{x}_k)$ as input, outputs a value close to 1, if $\mathbf{x}_i$ should be higher than $\mathbf{x}_k$ in the ranking; otherwise, f outputs a value close to 0. At the test time, the final ranking for an unlabeled example $\mathcal{X}$ is obtained by aggregating the predictions for all pairs of documents in $\mathcal{X}$. The pairwise approach works better than pointwise, but is still far from perfect.

The state of the art rank learning algorithms, such as **LambdaMART**, implement the listwise approach. In the listwise approach, we try to optimize the model directly on some metric that reflects the quality of ranking. There are various metrics for assessing search engine result ranking, including precision and recall. One popular metric that combines both precision and recall is called **mean average precision** (MAP).

To define MAP, let us ask judges (Google call those people *rankers*) to examine a collection of search results for a query and assign relevancy labels to each search result. Labels could be binary (1 for "relevant" and 0 for "irrelevant") or on some scale, say from 1 to 5: the higher the value, the more relevant the document is to the search query. Let our judges build such relevancy labeling for a collection of 100 queries. Now, let us test our ranking model on this collection. The **precision** of our model for some query is given by:

$$\text{precision} = \frac{|\{\text{relevant documents}\} \cap \{\text{retrieved documents}\}|}{|\{\text{retrieved documents}\}|},$$

where the notation $|\cdot|$ means "the number of." The **average precision** metric, AveP, is defined for a ranked collection of documents returned by a search engine for a query q as,

$$\text{AveP}(q) = \frac{\sum_{k=1}^{n}(P(k) \cdot \text{rel}(k))}{|\{\text{relevant documents}\}|},$$

where n is the number of retrieved documents, $P(k)$ denotes the precision computed for the top k search results returned by our ranking model for the query, $\text{rel}(k)$ is an indicator function equaling 1 if the item at rank k is a relevant document (according to judges) and zero otherwise. Finally, the MAP for a collection of search queries of size Q is given by,

$$\text{MAP} = \frac{\sum_{q=1}^{Q} \text{AveP}(q)}{Q}.$$

Now we get back to LambdaMART. This algorithm implements a listwise approach, and it uses gradient boosting to train the ranking function $h(\mathbf{x})$. Then, the binary model $f(\mathbf{x}_i, \mathbf{x}_k)$ that predicts whether the document $\mathbf{x}_i$ should have a higher rank than the document $\mathbf{x}_k$ (for the same search query) is given by a sigmoid with a hyperparameter α,

$$f(\mathbf{x}_i, \mathbf{x}_k) \overset{\text{def}}{=} \frac{1}{1 + \exp((h(\mathbf{x_i}) - h(\mathbf{x_k}))\alpha}.$$

Again, as with many models that predict probability, the cost function is cross-entropy computed using the model f. In our gradient boosting, we combine multiple regression trees to build the function h by trying to minimize the cost. Remember that in gradient boosting we add a tree to the model to reduce the error that the current model makes on the training data. For the classification problem, we computed the derivative of the cost function to replace real labels of training examples with these derivatives. LambdaMART works similarly, with one exception. It replaces the real gradient with a combination of the gradient and another factor that depends on the metric, such as MAP. This factor modifies the original gradient by increasing or decreasing it so that the metric value is improved.

That is a very bright idea and not many supervised learning algorithms can boast that they optimize a metric directly. Optimizing a metric is what we really want, but what we do in a typical supervised learning algorithm is we optimize the cost instead of the metric (because metrics are usually not differentiable). Usually, in supervised learning, as soon as we have found a model that optimizes the cost function, we try to tweak hyperparameters to improve the value of the metric. LambdaMART optimizes the metric directly.

The remaining question is how do we build the ranked list of results based on the predictions of the model f which predicts whether its first input has to be ranked higher than the second input. It's generally a computationally hard problem, and there are multiple implementations of rankers capable of transforming pairwise comparisons into a ranking list.

The most straightforward approach is to use an existing sorting algorithm. Sorting algorithms sort a collection of numbers in increasing or decreasing order. (The simplest sorting algorithm is called *bubble sort*. It's usually taught in engineering schools.) Typically, sorting algorithms iteratively compare a pair of numbers in the collection and change their positions in the list based on the result of that comparison. If we plug our function f into a sorting algorithm to execute this comparison, the sorting algorithm will sort documents and not numbers.

10.3   Learning to Recommend

Learning to recommend is an approach to building recommender systems. Usually, we have a user who consumes content. We have the history of consumption and want to suggest new content to this user that they would like. It could be a movie on Netflix or a book on Amazon.

Traditionally, two approaches were used to give recommendations: **content-based filtering** and **collaborative filtering**.

Content-based filtering consists of learning what users like based on the description of the content they consume. For example, if the user of a news site often reads news articles on science and technology, then we would suggest more documents on science and technology to this user. More generally, we could create one training set *per user* and add news articles to this dataset as a feature vector $\mathbf{x}$ and whether the user recently read this news article as a label y. Then we build the model of each user and can regularly examine each new piece of content to determine whether a specific user would read it or not.

The content-based approach has many limitations. For example, the user can be trapped in the so-called filter bubble: the system will always suggest to that user the information that looks very similar to what user already consumed. That could result in complete isolation of the user from information that disagrees with their viewpoints or expands them. On a more practical side, the users might just stop following recommendations, which is undesirable.

Collaborative filtering has a significant advantage over content-based filtering: the recommendations to one user are computed based on what other users consume or rate. For instance, if two users gave high ratings to the same ten movies, then it's more likely that user 1 will appreciate new movies recommended based on the tastes of the user 2 and vice versa. The drawback of this approach is that the content of the recommended items is ignored.

In collaborative filtering, the information on user preferences is organized in a matrix. Each row corresponds to a user, and each column corresponds to a piece of content that user rated or consumed. Usually, this matrix is huge and extremely sparse, which means that most of its cells aren't filled (or filled with a zero). The reason for such a sparsity is that most users consume or rate just a tiny fraction of available content items. It's is very hard to make meaningful recommendations based on such sparse data.

Most real-world recommender systems use a hybrid approach: they combine recommendations obtained by the content-based and collaborative filtering models.

I already mentioned that a content-based recommender model could be built using a classification or regression model that predicts whether a user will like the content based on the content's features. Examples of features could include the words in books or news articles the user liked, the price, the recency of the content, the identity of the content author and so on.

Two effective recommender system learning algorithms are **factorization machines** (FM) and **denoising autoencoders** (DAE).

10.3.1 Factorization Machines

Factorization machine is a relatively new kind of algorithm. It was explicitly designed for sparse datasets. Let's illustrate the problem.

In Figure 10.1 you see an example of sparse feature vectors with labels. Each feature vector represents information about one specific user and one specific movie. Features in the blue section represent a user. Users are encoded as one-hot vectors. Features in the green section

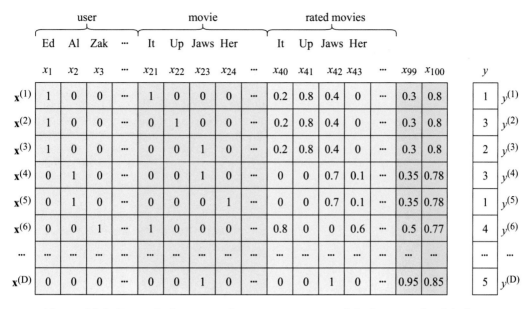

Figure 10.1: Example for sparse feature vectors $\mathbf{x}$ and their respective labels y.

represent a movie. Movies are also encoded as one-hot vectors. Features in the yellow section represent normalized scores the user in blue gave to each movie they rated. Feature x_{99} represents the ratio of movies with an Oscar among those the user has watched. Feature x_{100} represents the percentage of the movie watched by the user in blue before they scored the movie in green. The target y is the score given by the user in blue to the movie in green.

Real recommender systems can have millions of users, so the matrix in Figure 10.1 can count hundreds of millions of rows. The number of features could also be millions, depending on how rich is the choice of content is and how creative you, as a data analyst, are in feature engineering. Features x_{99} and x_{100} were handcrafted during the feature engineering process, and I only show two features for the purposes of illustration.

Trying to fit a regression or classification model to such an extremely sparse dataset would result in poor generalization. Factorization machines approach this problem differently.

The factorization machine model is defined as follows:

$$f(\mathbf{x}) \overset{\text{def}}{=} b + \sum_{i=1}^{D} w_i x_i + \sum_{i=1}^{D} \sum_{j=i+1}^{D} (\mathbf{v}_i \mathbf{v}_j) x_i x_j,$$

where b and w_i, $i = 1, \ldots, D$, are scalar parameters similar to those used in linear regression. Vectors $\mathbf{v}_i$ are k-dimensional vectors of **factors**. k is a hyperparameter and is usually much smaller than D. The expression $\mathbf{v}_i \mathbf{v}_j$ is a dot-product of the ith and j^{th} vectors of factors. As you can see, instead looking for one wide vector of parameters, which can reflect interactions between features poorly because of sparsity, we complete it by additional parameters that apply to pairwise interactions $x_i x_j$ between features. However, instead of having a parameter $w_{i,j}$ for each interaction, which would add an enormous[1] quantity of new parameters to the model, we factorize $w_{i,j}$ into $\mathbf{v}_i \mathbf{v}_j$ by adding only $Dk \ll D(D-1)$ parameters to the model[2].

Depending on the problem, the loss function could be squared error loss (for regression) or hinge loss. For classification with $y \in \{-1, +1\}$, with hinge loss or logistic loss the prediction is made as $y = \text{sign}(f(x))$. The logistic loss is defined as,

$$loss(f(\mathbf{x}), y) = \frac{1}{\ln 2} \ln(1 + e^{-y f(\mathbf{x})}).$$

Gradient descent can be used to optimize the average loss. In the example in Figure 10.1, the labels are in $\{1, 2, 3, 4, 5\}$, so it's a multiclass problem. We can use the **one versus rest** strategy to convert this multiclass problem into five binary classification problems.

10.3.2 Denoising Autoencoders

From Chapter 7, you know what a **denoising autoencoder** is: it's a neural network that reconstructs its input from the bottleneck layer. The fact that the input is corrupted by noise while the output shouldn't be makes denoising autoencoders an ideal tool to build a recommender model.

The idea is very straightforward: new movies a user could like are seen as if they were removed from the complete set of preferred movies by some corruption process. The goal of the denoising autoencoder is to reconstruct those removed items.

To prepare the training set for our denoising autoencoder, remove the blue and green features from the training set in Figure 10.1. Because now some examples become duplicates, keep only the unique ones.

At the training time, randomly replace some of the non-zero yellow features in the input feature vectors with zeros. Train the autoencoder to reconstruct the uncorrupted input.

At prediction time, build a feature vector for the user. The feature vector will include uncorrupted yellow features as well as the handcrafted features like x_{99} and x_{100}. Use the trained DAE model to reconstruct the uncorrupted input. Recommend to the user movies that have the highest scores at the model's output.

Another effective collaborative-filtering model is an FFNN with two inputs and one output. Remember from Chapter 8 that neural networks are good at

---

[1]To be more precise we would add $D(D-1)$ parameters $w_{i,j}$.
[2]The notation $\ll$ means "much less than."

handling multiple simultaneous inputs. A training example here is a triplet $(\mathbf{u}, \mathbf{m}, r)$. The input vector $\mathbf{u}$ is a **one-hot encoding** of a user. The second input vector $\mathbf{m}$ is a one-hot encoding of a movie. The output layer could be either a sigmoid (in which case the label r is in $[0, 1]$) or ReLU, in which case r can be in some typical range, $[1, 5]$ for example.

10.4   Self-Supervised Learning: Word Embeddings

We have already discussed word embeddings in Chapter 7. Recall that **word embeddings** are feature vectors that represent words. They have the property that similar words have similar feature vectors. The question that you probably wanted to ask is where these word embeddings come from. The answer is (again): they are learned from data.

There are many algorithms to learn word embeddings. Here, we consider only one of them: **word2vec**, and only one version of word2vec called **skip-gram**, which works well in practice. Pretrained word2vec embeddings for many languages are available to download online.

In word embedding learning, our goal is to build a model which we can use to convert a one-hot encoding of a word into a word embedding. Let our dictionary contain 10,000 words. The one-hot vector for each word is a 10,000-dimensional vector of all zeroes except for one dimension that contains a 1. Different words have a 1 in different dimensions.

Consider a sentence: "I almost finished reading the book on machine learning." Now, consider the same sentence from which we have removed one word, say "book." Our sentence becomes: "I almost finished reading the · on machine learning." Now let's only keep the three words before the · and three words after: "finished reading the · on machine learning." Looking at this seven-word window around the ·, if I ask you to guess what · stands for, you would probably say: "book," "article," or "paper." That's how the context words let you predict the word they surround. It's also how the machine can learn that words "book," "paper," and "article" have a similar meaning: because they share similar contexts in multiple texts.

It turns out that it works the other way around too: a word can predict the context that surrounds it. The piece "finished reading the · on machine learning" is called a skip-gram with window size 7 (3 + 1 + 3). By using the documents available on the Web, we can easily create hundreds of millions of skip-grams.

Let's denote a skip-gram like this: $[\mathbf{x}_{-3}, \mathbf{x}_{-2}, \mathbf{x}_{-1}, \mathbf{x}, \mathbf{x}_{+1}, \mathbf{x}_{+2}, \mathbf{x}_{+3}]$. In our sentence, $\mathbf{x}_{-3}$ is the one-hot vector for "finished," $\mathbf{x}_{-2}$ corresponds to "reading," $\mathbf{x}$ is the skipped word (·), $\mathbf{x}_{+1}$ is "on" and so on. A skip-gram with window size 5 will look like this: $[\mathbf{x}_{-2}, \mathbf{x}_{-1}, \mathbf{x}, \mathbf{x}_{+1}, \mathbf{x}_{+2}]$.

The skip-gram model with window size 5 is schematically depicted in Figure 10.2. It is a fully-connected network, like the multilayer perceptron. The input word is the one denoted as · in the skip-gram. The neural network has to learn to predict the context words of the skip-gram given the central word.

You can see now why the learning of this kind is called **self-supervised**: the labeled examples get extracted from the unlabeled data such as text.

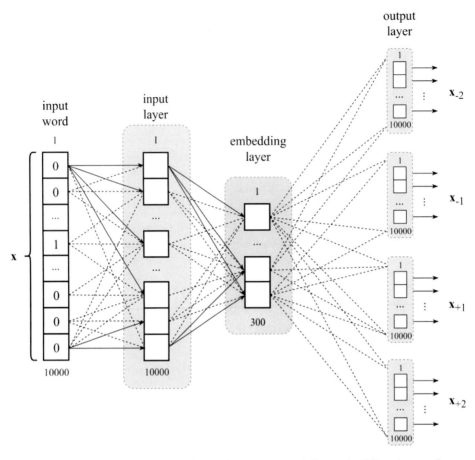

Figure 10.2: The skip-gram model with window size 5 and the embedding layer of 300 units.

The activation function used in the output layer is softmax. The cost function is the negative log-likelihood. The embedding for a word is obtained as the output of the embedding layer when the one-hot encoding of this word is given as the input to the model.

Because of the large number of parameters in the word2vec models, two techniques are used to make the computation more efficient: *hierarchical softmax* (an efficient way of computing softmax that consists in representing the outputs of softmax as leaves of a binary tree) and *negative sampling* (where the idea is only to update a random sample of all outputs per iteration of gradient descent). I leave these for further reading.

Chapter 11

Conclusion

Wow, that was fast! You are really good if you got here and managed to understand most of the book's material.

If you look at the number at the bottom of this page, you see that I have overspent paper, which means that the title of the book was slightly misleading. I hope that you forgive me for this little marketing trick. After all, if I wanted to make this book exactly a hundred pages, I could reduce font size, white margins, and line spacing, or remove the section on UMAP and leave you on your own with the original paper. Believe me: you would not want to be left on your own with the original paper on UMAP! (Just kidding.)

However, by stopping now, I feel confident that you have got everything you need to become a great modern data analyst or machine learning engineer. That doesn't mean that I covered everything, but what I covered in a hundred+ pages you would find in a bunch of books, each a thousand pages thick. Much of what I covered is not in the books at all: typical machine learning books are conservative and academic, while I emphasized those algorithms and methods that you will find useful in your day to day work.

What exactly would I have covered if it was a thousand-page machine learning book?

11.1   What Wasn't Covered

11.1.1   Topic Modeling

In text analysis, topic modeling is a prevalent unsupervised learning problem. You have a collection of text documents, and you would like to discover topics present in each document. **Latent Dirichlet Allocation** (LDA) is a very effective algorithm of topic discovery. You decide how many topics are present in your collection of documents and the algorithm assigns a topic to each word in this collection. Then, to extract the topics from a document, you simply count how many words of each topic are present in that document.

11.1.2 Gaussian Processes

Gaussian processes (GP) is a supervised learning method that competes with kernel regression. It has some advantages over the latter. For example, it provides confidence intervals for the regression line in each point. I decided not to explain GP because I could not figure out a simple way to explain them, but you definitely could spend some time to learn about GP. It will be time well spent.

11.1.3 Generalized Linear Models

Generalized Linear Model (GLM) is a generalization of the linear regression to modeling various forms of dependency between the input feature vector and the target. Logistic regression, for instance, is one form of GLMs. If you are interested in regression and you look for simple and explainable models, you should definitely read more on GLM.

11.1.4 Probabilistic Graphical Models

I have mentioned one example of **probabilistic graphical models** (PGMs) in Chapter 7: **conditional random fields** (CRF). With CRF you can model the input sequence of words and relationships between the features and labels in this sequence as a sequential *dependency graph*. More generally, a PGM can be any graph. A **graph** is a structure consisting of a collection of nodes and edges that each join a pair of nodes. Each node in PGM represents some random variable (values of which can be observed or unobserved), and edges represent the conditional dependence of one random variable on another random variable. For example, the random variable "sidewalk wetness" depends on the random variable "weather condition." By observing values of some random variables, an optimization algorithm can learn from data the dependency between observed and unobserved variables.

If you decide to learn more about PGMs, they are also known as Bayesian networks, belief networks, and probabilistic independence networks.

11.1.5 Markov Chain Monte Carlo

If you work with graphical models and want to sample examples from a very complex distribution defined by the dependency graph, you could use **Markov Chain Monte Carlo** (MCMC) algorithms. MCMC is a class of algorithms for sampling from any probability distribution defined mathematically. Remember that when we talked about **denoising autoencoders**, we sampled noise from the normal distribution. Sampling from standard distributions, such as normal or uniform, is relatively easy because their properties are well known. However, the task of sampling becomes significantly more complicated when the probability distribution can have an arbitrary form defined by a complex formula.

11.1.6   Generative Adversarial Networks

Generative adversarial networks, or GANs, are a class of neural networks used in unsupervised learning. They are implemented as a system of two neural networks contesting with each other in a *zero-sum game* setting. The most popular application of GANs is to learn to generate photographs that look authentic to human observers. The first of the two networks takes a random input (typically Gaussian noise) and learns to generate an image as a matrix of pixels. The second network takes as input two images: one "real" image from some collection of images as well as the image generated by the first network. The second network has to learn to recognize which one of the two images was generated by the first network. The first network gets a negative loss if the second network recognizes the "fake" image. The second network, on the other hand, gets penalized if it fails to recognize which one of the two images is fake.

11.1.7   Genetic Algorithms

Genetic algorithms (GA) are a numerical optimization technique used to optimize undifferentiable optimization objective functions. They use concepts from evolutionary biology to search for a global optimum (minimum or maximum) of an optimization problem by mimicking evolutionary biological processes.

GA work by starting with an initial generation of candidate solutions. If we look for optimal values of the parameters of our model, we first randomly generate multiple combinations of parameter values. We then test each combination of parameter values against the objective function. Imagine each combination of parameter values as a point in a multi-dimensional space. We then generate a subsequent generation of points from the previous generation by applying such concepts as "selection," "crossover," and "mutation."

In a nutshell, that results in each new generation keeping more points similar to those points from the previous generation that performed the best against the objective. In the new generation, the points that performed the worst in the previous generation are replaced by "mutations" and "crossovers" of the points that performed the best. A mutation of a point is obtained by a random distortion of some attributes of the original point. A crossover is a certain combination of several points (for example, an average).

Genetic algorithms allow the finding of solutions to any measurable optimization criteria. For example, GA can be used to optimize the hyperparameters of a learning algorithm. They are typically much slower than gradient-based optimization techniques.

11.1.8 Reinforcement Learning

As we already discussed, **reinforcement learning** (RL) solves a very specific kind of problem where the decision making is sequential. Usually, there's an agent acting in an unknown environment. Each action brings a reward and moves the agent to another state of the environment (usually, as a result of some random process with unknown properties). The goal of the agent is to optimize its long-term reward.

Reinforcement learning algorithms, such as Q-learning, and their neural network based counterparts are used in learning to play video games, robotic navigation and coordination, inventory and supply chain management, optimization of complex electric power systems (power grids), and the learning of financial trading strategies.

$$* * *$$

The book stops here. Don't forget to occasionally visit the book's companion wiki to stay updated on new developments in each machine learning area considered in the book. As I said in the Preface, this book, thanks to the constantly updated wiki, like a good wine keeps getting better after you buy it.

Oh, and don't forget that the book is distributed on the *read first, buy later* principle. That means that if while reading these words you look at a PDF file and cannot remember having paid to get it, you are probably the right person for buying the book.

11.2 Acknowledgements

The high quality of this book would be impossible without volunteering editors. I especially thank the following readers for their systematic contributions: Bob DuCharme, Martijn van Attekum, Daniel Maraini, Ali Aziz, Rachel Mak, Kelvin Sundli, and John Robinson.

Other wonderful people to whom I am grateful for their help are Michael Anuzis, Knut Sverdrup, Freddy Drennan, Carl W. Handlin, Abhijit Kumar, Lasse Vetter, Ricardo Reis, Daniel Gross, Johann Faouzi, Akash Agrawal, Nathanael Weill, Filip Jekic, Abhishek Babuji, Luan Vieira, Sayak Paul, Vaheid Wallets, Lorenzo Buffoni, Eli Friedman, Łukasz Mądry, Haolan Qin, Bibek Behera, Jennifer Cooper, Nishant Tyagi, Denis Akhiyarov, Aron Janarv, Alexander Ovcharenko, Ricardo Rios, Michael Mullen, Matthew Edwards, David Etlin, Manoj Balaji J, David Roy, Luan Vieira, Luiz Felix, Anand Mohan, Hadi Sotudeh, Charlie Newey, Zamir Akimbekov, Jesus Renero, Karan Gadiya, Mustafa Anıl Derbent, JQ Veenstra, Zsolt Kreisz, Ian Kelly, Lukasz Zawada, Robert Wareham, Thomas Bosman, Lv Steven, Ariel Rossanigo, Michael Lumpkins, Secil Sozuer, Boris Kouambo, Yi Jayeon, Tim Flocke, and Luciano Segura.

Index

graph, 27, 134
grid search, 59
GRU, 74

HDBSCAN, 112
hinge loss, 31
hyperparameter tuning, 59

ID3, 27
intersection, 10
interval, 13
  open, 13

k-means, 81, 110
k-Nearest Neighbors, 19, 34
kernel, 6, 33, 77
  RBF, 33
kernel trick, 32

label, 2, 19
LambdaMART, 126
Latent Dirichlet Allocation, 133
layer, 20, 61, 62
  bottleneck, 92, 118
  hidden, 65
LDA, 133
learning
  unsupervised, 135
  active, 89
  deep, 20, 65
  ensemble, 83
  listwise, 125
  metric, 123
  one-shot, 94, 125
  pairwise, 125
  pointwise, 125
  reinforcement, 3, 136
  self-, 91
  self-supervised, 131
  semi-supervised, 2, 91
  seq2seq, 88
  sequence-to-sequence, 88
  shallow, 20, 101
  supervised, 1

  transfer, 102
  unsupervised, 2, 107
  zero-shot, 95
learning algorithm, 4
  classification, 19
  incremental, 48
  regression, 19
  semi-supervised, 2
  supervised, 2
  unsupervised, 2
learning to rank, 125
learning to recommend, 127
leave one out estimate, 109
likelihood, 26
log-likelihood, 27
long short-term memory, 74
LSTM, 74

majority vote, 99
MAP, 18, 126
margin, 5
Markov Chain Monte Carlo, 134
matrix, 9
max, 13
maximum a posteriori, 18
maximum likelihood, 26, 80, 115
MCMC, 134
mean, 16
mean integrated squared error, 108
mean model, 54
mean squared error, 26
meta-model, 83
min, 13
minibatch stochastic gradient descent, 40
minimal gated GRU, 75
minimum
  global, 13
  local, 13
MLP, 62
MND, 79
model, 2, 4, 5, 19
  nonparametric, 28, 107
  parametric, 28, 107
  sparse, 53

50793059R00088

Made in the USA
Middletown, DE
28 June 2019